Lecture Notes in Computer Science 16080

Founding Editors

Gerhard Goos
Juris Hartmanis

Editorial Board Members

Elisa Bertino, *Purdue University, West Lafayette, IN, USA*
Wen Gao, *Peking University, Beijing, China*
Bernhard Steffen ⓘ, *TU Dortmund University, Dortmund, Germany*
Moti Yung ⓘ, *Columbia University, New York, NY, USA*

The series Lecture Notes in Computer Science (LNCS), including its subseries Lecture Notes in Artificial Intelligence (LNAI) and Lecture Notes in Bioinformatics (LNBI), has established itself as a medium for the publication of new developments in computer science and information technology research, teaching, and education.

LNCS enjoys close cooperation with the computer science R & D community, the series counts many renowned academics among its volume editors and paper authors, and collaborates with prestigious societies. Its mission is to serve this international community by providing an invaluable service, mainly focused on the publication of conference and workshop proceedings and postproceedings. LNCS commenced publication in 1973.

Abdelkader Hameurlain · A Min Tjoa
Editors

Transactions on Large-Scale Data- and Knowledge-Centered Systems LVIII

Springer

Editors-in-Chief
Abdelkader Hameurlain
IRIT, Paul Sabatier University
Toulouse, France

A Min Tjoa
Faculty of Informatics, TU WIEN (Vienna
University of Technology)
Vienna, Austria

ISSN 0302-9743 ISSN 1611-3349 (electronic)
Lecture Notes in Computer Science
ISSN 1869-1994 ISSN 2510-4942 (electronic)
Transactions on Large-Scale Data- and Knowledge-Centered Systems
ISBN 978-3-662-72115-5 ISBN 978-3-662-72116-2 (eBook)
https://doi.org/10.1007/978-3-662-72116-2

© The Editor(s) (if applicable) and The Author(s), under exclusive license
to Springer-Verlag GmbH, DE, part of Springer Nature 2026

This work is subject to copyright. All rights are solely and exclusively licensed by the Publisher, whether the whole or part of the material is concerned, specifically the rights of translation, reprinting, reuse of illustrations, recitation, broadcasting, reproduction on microfilms or in any other physical way, and transmission or information storage and retrieval, electronic adaptation, computer software, or by similar or dissimilar methodology now known or hereafter developed.
The use of general descriptive names, registered names, trademarks, service marks, etc. in this publication does not imply, even in the absence of a specific statement, that such names are exempt from the relevant protective laws and regulations and therefore free for general use.
The publisher, the authors and the editors are safe to assume that the advice and information in this book are believed to be true and accurate at the date of publication. Neither the publisher nor the authors or the editors give a warranty, expressed or implied, with respect to the material contained herein or for any errors or omissions that may have been made. The publisher remains neutral with regard to jurisdictional claims in published maps and institutional affiliations.

This Springer imprint is published by the registered company Springer-Verlag GmbH, DE,
part of Springer Nature.
The registered company address is: Heidelberger Platz 3, 14197 Berlin, Germany

If disposing of this product, please recycle the paper.

Preface

This volume contains five fully revised regular papers, covering a wide range of very hot topics focused on storing and querying evolving graphs in NoSQL storage models, the data assetization journey, deep learning rainfall nowcasts, fairness assessment, and energy and flexibility based on FlexOffers and Blockchain.

We would like to sincerely thank the editorial board for thoroughly refereeing the submitted papers and ensuring the high quality of this volume. In addition, we would like to express our wholehearted thanks to the team at Springer for their ready availability, the efficiency of their management and the very pleasant cooperation in the realization of the TLDKS journal volumes.

July 2025

Abdelkader Hameurlain
A Min Tjoa

Organization

Editors-in-Chief

Abdelkader Hameurlain Toulouse University, IRIT, France
A Min Tjoa TU WIEN (Vienna University of Technology)

Editorial Board

Reza Akbarinia Inria, France
Dagmar Auer Johannes Kepler University Linz, Austria
Djamal Benslimane University Lyon 1, France
Mirel Cosulschi University of Craiova, Romania
Johann Eder Alpen Adria University of Klagenfurt, Austria
Anna Formica National Research Council in Rome, Italy
Shahram Ghandeharizadeh University of Southern California, USA
Anastasios Gounaris Aristotle University of Thessaloniki, Greece
Sergio Ilarri University of Zaragoza, Spain
Petar Jovanovic Universitat Politècnica de Catalunya and BarcelonaTech, Spain
Aida Kamišalić Latifić University of Maribor, Slovenia
Dieter Kranzlmüller Ludwig-Maximilians-Universität München, Germany
Philippe Lamarre INSA Lyon, France
Lenka Lhotská Technical University of Prague, Czech Republic
Vladimir Marik Technical University of Prague, Czech Republic
Jorge Martinez Gil Software Competence Center Hagenberg, Austria
Riad Mokadem Toulouse University, IRIT, France
Franck Morvan Toulouse University, IRIT, France
Torben Bach Pedersen Aalborg University, Denmark
Günther Pernul University of Regensburg, Germany
Viera Rozinajova Kempelen Institute of Intelligent Technologies, Slovakia
Soror Sahri LIPADE, Université Paris Cité, France
Joseph Vella University of Malta, Malta
Shaoyi Yin Toulouse University, IRIT, France
Feng "George" Yu Youngstown State University, USA

Contents

Storing and Querying Evolving Graphs in NoSQL Storage Models 1
   Alexandros Spitalas, Anastasios Gounaris, Andreas Kosmatopoulos,
   and Kostas Tsichlas

Data Assetization Journey: Concepts, Principles, and Illustrations 45
   Zakaria Maamar, Belkacem Chikhaoui, Amel Benna, Vanilson Burégio,
   and Djamal Benslimane

Do Echo Top Heights Improve Deep Learning Rainfall Nowcasts? A Case
Study in the Netherlands 66
   Peter Pavlík, Marc Schleiss, Anna Bou Ezzeddine, and Viera Rozinajová

A Deep Dive into FAIRness Assessment: UReFM, a Formal Framework
for Representing, Analyzing and Comparing Measures 93
   Philippe Lamarre, Jennie Andersen, Alban Gaignard, and Sylvie Cazalens

Jointly Trading Energy and Flexibility Based on FlexOffers and Blockchain ... 128
   Laurynas Siksnys and Torben Bach Pedersen

Author Index 161

Storing and Querying Evolving Graphs in NoSQL Storage Models

Alexandros Spitalas[1], Anastasios Gounaris[2(✉)], Andreas Kosmatopoulos[2], and Kostas Tsichlas[1]

[1] University of Patras, Patras, Greece
{a.spitalas,ktsichlas}@ceid.upatras.gr
[2] Aristotle University of Thessaloniki, Thessaloniki, Greece
{gounaria,akosmato}@csd.auth.gr

Abstract. This paper investigates advanced storage models for evolving graphs, focusing on the efficient management of historical data and the optimization of global query performance. Evolving graphs, which represent dynamic relationships between entities over time, present unique challenges in preserving their complete history while supporting complex analytical queries. We first do a fast review of the current state of the art focusing mainly on distributed historical graph databases to provide the context of our proposals. We investigate the implementation of an enhanced vertex-centric storage model in MongoDB that prioritizes space efficiency by leveraging in-database query mechanisms to minimize redundant data and reduce storage costs. To ensure broad applicability, we employ datasets, some of which are generated with the LDBC SNB generator, appropriately post-processed to utilize both snapshot- and interval-based representations. Our experimental results both in centralized and distributed infrastructures, demonstrate significant improvements in query performance, particularly for resource-intensive global queries that traditionally suffer from inefficiencies in entity-centric frameworks. The proposed model achieves these gains by optimizing memory usage, reducing client involvement, and exploiting the computational capabilities of MongoDB. By addressing key bottlenecks in the storage and processing of evolving graphs, this study demonstrates a step toward a robust and scalable framework for managing dynamic graph data. This work contributes to the growing field of temporal graph analytics by enabling more efficient exploration of historical data and facilitating real-time insights into the evolution of complex networks.

1 Introduction

Consider the challenge of tracking the spread of an infectious disease through global air travel. To understand how an outbreak evolved, we need to analyze how passengers moved between cities over time, identifying critical points where the disease was likely transmitted. A static representation of the flight network is insufficient, as routes change daily, restrictions alter travel patterns, and the contagion follows a dynamic trajectory. Efficiently storing and querying this evolving network is essential to reconstruct past transmission pathways, predict future outbreaks, and inform containment strategies. The ability to retrieve historical snapshots at any time granularity and analyze evolving connectivity patterns is crucial for timely decision-making, highlighting the need for scalable and optimized temporal graph storage solutions.

The analysis of networks that evolve over time has gained increasing attention across multiple scientific disciplines, including sociology [13], physics [56], ecology [48], computer science [43], and engineering [54]. These networks - commonly referred to as dynamic, adaptive, time-varying, evolving, or temporal networks, depending partly on the scientific area under consideration - capture interactions between entities that change over time. While different fields emphasize distinct aspects of temporal graphs, a common challenge remains: representing and querying the historical evolution of such networks efficiently. Understanding how relationships form, persist, and dissolve is critical for applications ranging from social influence modeling to infrastructure resilience analysis.

To provide a unified framework for studying these networks, the concept of Time-Varying Graphs (TVGs) [8] has been introduced, offering a formalism to express various temporal network models. This formalism represents entities as nodes and their relationships as edges, both of which can be annotated with attributes (such as name and weight). TVGs facilitate the application of time-aware algorithms, enabling tasks such as temporal shortest paths, influence spread analysis, and anomaly detection in dynamic systems. However, while substantial progress has been made in defining the theoretical foundations and computational techniques for temporal graphs, less attention has been given to the underlying storage models that support these operations. Without efficient storage solutions, retrieving historical snapshots, executing time-dependent queries, and scaling to large datasets remain computationally expensive and impractical. Indeed, many traditional database systems struggle to balance storage efficiency with query performance, especially for large-scale temporal graphs. This gap necessitates the development of storage models that can compactly represent historical data while enabling fast retrieval of both localized and global queries.

To give an example, consider a graph corresponding to a social network that comprises millions of users (vertices) and the friendship relationships between them (edges). By examining the graph through two subsequent days, we witness a number of newly created, removed, and altered vertices or edges and a significant fraction of the graph that has remained the same across the two days. A system that aims to efficiently store all the snapshots of such a graph should employ techniques that mitigate the presence of unaltered data between different snapshots (i.e., take advantage of the commonalities between snapshots and refrain from storing duplicate data across snapshots). Moreover, the consideration of snapshots results in constraining the time granularity of the graph. Two consecutive snapshots could be defined over two consecutive years, months, days, hours, minutes, etc. This means that in general, snapshots are merely a view of the network at a particular time granularity and could change based on the user-defined query. However, in many systems, for performance reasons, the time granularity is predefined and cannot change since it requires the partial or total rebuilding of the system.

In a nutshell, this paper has a three-fold contribution; 1) it reviews the current research on historical graph management systems, 2) it evaluates extensively a vertex-centric approach that leverages MongoDB, and 3) it identifies a lack of temporal graph generators and makes a first step toward this direction by creating historical graph datasets based on the LDBC SNB [3].[1]

The survey of existing storage techniques for temporal graphs, categorizes prior works based on their system architecture (distributed vs. non-distributed) focusing on distributed systems. In addition, it highlights trade-offs between different approaches and identifies key limitations in current solutions.

We also explore a vertex-centric storage model that leverages MongoDB to enhance the efficiency of temporal graph storage and retrieval. Unlike traditional snapshot-based approaches, this model optimizes space utilization by structuring historical data in a vertex-centric manner, reducing redundancy while maintaining fast access to both local and global queries. We compare our results to a similar Cassandra-based implementation [31]. Overall, the MongoDB implementation can perform more complex in-database queries and decrease client involvement in query processing. Additionally, instead of getting the documents from the database in a single large batch, we explore the option to employ a `foreach` approach (when this is expected to be more efficient), and as a result, to further reduce client involvement and memory requirements.

Finally, to validate our approach we use/generate historical graph datasets and conduct extensive experiments to explore the performance of different versions of our storage model in various settings. These settings include transactions or graph-scale operations, batch or streaming modes, and the use of snapshots or time intervals. Our evaluation measures query execution times, storage efficiency, and scalability, with a focus on complex global queries that are traditionally expensive in entity-centric systems. The results both in a centralized as well as in a distributed setting, indicate significant improvements in execution speed, reduced memory consumption, and lower client-side processing overhead, allowing us to execute more expensive queries in the same infrastructure. To generate some of the historical graph datasets we made some progress on creating a generator based on the LDBC SNB generator.

The structure of the paper is as follows. In Sect. 2 we review the literature on distributed graph database systems that store the complete or the partial history of a historical graph. In Sect. 3, we describe the vertex-centric storage model given by some of the authors in [32], and provide details for implementation approaches of the vertex-centric schema in Cassandra as described in [31]. We describe our approach using MongoDB, which better exploits in-database query processing mechanisms in Sect. 4. Our proposal is thoroughly evaluated in Sect. 5

---

[1] This paper partially contains heavily updated material from previously published conference papers by the same authors [58,59]. In particular, apart from the new material explicitly stated in Sect. ??, all experiments have been re-conducted in different settings (local and cluster modes).

both in a centralized and in a distributed environment, and finally, we conclude in Sect. 6.

2 Background and Related Work

There have been two main approaches regarding a TVG system's design [26,28, 47], the time-centric approach and the entity-centric approach. In the former case, the system is indexed according to the time instants (i.e., changes are organized by the time instant they occur), while in the latter case, the system is indexed according to the entities, their relationships, and their respective history (i.e. changes are organized based on the vertex or edge they refer to). Most of the previous research work aims at storing the changes themselves (known as deltas) that occur between different snapshots. A system that maintains sets of deltas is thus able to reconstruct any particular snapshot by sequentially applying all the deltas up to the desired time instant. The above-described kind of framework can be used in both approaches but lends itself more naturally to a time-centric approach.

Entity-centric approaches can be separated into different groups of categories, with the most common being vertex-centric and edge-centric. In both approaches, the idea is to store the whole history of an entity (vertex or edge) in one entity, taking advantage of temporal locality and space efficiency by reducing duplicate information. In this case, for the reconstruction of a snapshot, access to every entity of the graph is required, which can be time-consuming. However, by taking advantage of locality and space efficiency, entity-centric approaches can reach if not surpass the efficiency of time-centric approaches. The advantage of entity-centric approaches is mostly seen in local and global queries when instead of looking at a single time instant (a single snapshot), the query involves a time interval (a series of snapshots).

Another viewpoint concerning a system's design is based on the type of queries that the system should be able to evaluate. Local queries involve a particular vertex or a limited selection of adjacent vertices (e.g., the 2-hop neighborhood of a vertex). On the other hand, global queries consider the majority or the entirety of a graph's vertices (e.g., global clustering coefficient). Furthermore, both local and global queries should be able to be executed on either a time instant or on a range of time instants (e.g., average shortest path length between two vertices in the ten first snapshots). In the first case, a query aims to evaluate a measure at a particular time instant (e.g., shortest path length between two vertices at a particular time instant), while in the latter case, the query's objective is to extract information regarding a measure's evolution through snapshots for analytic purposes. By taking advantage of locality, entity-centric approaches will always be more efficient on local queries. However, regarding global queries, the efficiency depends on multiple factors, among which are the number of snapshots in the query and the specifics of the query.

Due to the rapid growth of historical graphs, storage efficiency is of great importance. Time-centric approaches have to store logs of events instead of copying snapshots in every instant, to improve storage efficiency. However, depending on the occasion, different techniques can be more beneficial; for example, in the extreme case of changing completely all entities at every time instant, simply storing from scratch each snapshot is probably the best idea regarding space efficiency. Entity-centric approaches use logs of events within each entity. This is why, entity-centric approaches can be space-optimal and can probably reach a better combination of time and storage efficiency.

In the early approaches before 2017, there were two main research directions regarding evolving graph storage processing. Systems for non-evolving graphs, such as Trinity [55], could be leveraged to support historical queries by explicitly storing each snapshot, but apparently, such solutions are inefficient. A comprehensive survey regarding these approaches for evolving graph data management can be found in [30] with the most notable proposals being those in [26,34,50,57]. In general, these techniques rely on the storage of snapshots and deltas (logging), which exhibits a trade-off between space and time. Having a large number of snapshots results in deltas of small size but the space cost is substantial since we need to maintain many copies of the graph. On the other hand, having a handful of snapshots means that deltas will be quite large, and queries at specific time instants may require a long time to execute. Three of these proposals operate in a parallel or distributed setting, i.e., DeltaGraph [26], TGI [28] and G^* [34]. The primary focus of DeltaGraph is the storage and retrieval of snapshots from an evolving graph sequence. It is best visualized as a rooted tree-like hierarchical graph structure with its leaves corresponding to snapshots of the sequence and the inner nodes corresponding to graphs that can be obtained by applying a differential function (e.g., intersection) to its children. DeltaGraph supports time point (single-point) queries, time interval snapshot queries, and multiple time point (multipoint) queries. Furthermore, along with the graph structure, a query is also able to return the attributes of vertices and edges (e.g., name or weight). The TGI system extends the functionality of the DeltaGraph system by providing support for operations that are concerned with individual vertices or neighborhoods. The G^* parallel system takes advantage of the commonalities that exist between snapshots by only storing each version of a vertex once and by avoiding storing redundant information that is not modified between different snapshots. Furthermore, G^* achieves substantial data locality since each G^* server is assigned its own set of vertices and corresponding entities. On the other hand, G^* uses some form of logging to store connection information between different entities.

Following the year 2017, a considerable amount of research has been conducted on historical graph management systems. A significant portion of these studies concentrates on enhancing the efficacy of main memory or streaming applications, as historical graphs tend to undergo frequent updates (e.g., [22,33]). In contrast to earlier years, a greater number of studies now focus on entity-centric approaches and techniques that minimize the spatial complexity of the system. Furthermore, while many researchers employ an existing storage back-end to implement their approach, some leverage a customized storage back-end to optimize storage and indexing. Not all indexes can effectively support the queries frequently used in historical graphs, such as spatial or range queries.

It is worth noting that significant advancements have been made towards Online Transaction Processing (OLTP) and Online Analytical Processing (OLAP) systems for historical graph database systems. Many applications that involve historical graphs entail concurrent transaction processing, which requires careful attention. Additionally, it is imperative for such systems to not only store data but also execute a wide range of analytical tasks, including those specialized for historical graphs such as querying the graph's history. Despite these advancements, we have a long way to go before reaching a point where such systems will be able to efficiently handle both OLTP and OLAP queries since their requirements to achieve efficiency point toward different directions.

2.1 Non-distributed Systems

In Table 1, we provide a list of non-distributed systems for historical graph management. We do not elaborate on these systems since we focus on distributed systems. In Table 2, we list all (to the best of our knowledge) distributed systems for historical graph management after the year 2017 with a few earlier systems as well. Since we focus on large Historical Graphs, where storage efficiency is important, it makes sense that our setting falls under the distributed category, so we discuss briefly some of these systems.

2.2 Distributed Systems

HiNode. This was the first storage model that adopted a pure vertex-centric approach. It was introduced in [32] and supports valid time as well as extensions like multiple universes. It was implemented within the G^* system [34] by replacing its storage subsystem. They showed gains in space usage, which is an immediate consequence of the pure vertex-centric approach. They supported local queries (e.g., 2-hop queries) as well as global queries (e.g., clustering coefficient). In addition, this vertex-centric model was also adapted for NoSQL databases by creating two models, SingleTable (ST) and MultipleTable (MT). In the former, all data fit in one table and a row has the data of a diachronic node, while in the latter, data are split into different tables (thus, slightly violating our pure vertex-centric approach). Two implementations were made, one in Cassandra [31] and later one in MongoDB [58] for comparison reasons. In MongoDB, indices and iterative computation were used for efficiency and to reduce memory usage.

Portal. In [44,46], the authors discuss interval-based models (where time is represented by intervals) and point-based models (where time is represented by a sequence of time instants) for time queries, focusing on the interval-based model with sequenced semantics. They propose a Temporal Graph Algebra (TGA) and a temporal graph model (TGraph) supporting TGA. In addition, in [45] they propose a declarative language (Portal) based on the previous model and built on top of a distributed system (Apache Spark). The Portal has SQL-like syntax that

Table 1. Non-distributed systems for historical graph management. The second column ("Memory") shows whether the system is implemented in main or external memory. The third column ("Storage Model") shows whether they use a custom-based model or an already implemented one (like Neo4j). The third column ("Time-related characteristics") shows the assumptions regarding the notion of time.

Summarizing the Characteristics of Non-Distributed Temporal Graph Management Systems			
Systems	Memory	Storage Model	Time-related characteristics
InteractionGraph [15]	Main Memory (old graph in disk)	Custom	Transaction time
STVG [40]	Main Memory	Neo4j	Valid time, offline, restricted to transit networks
ASPEN [10]	In-Memory/parallel	extends Ligra	Streaming
GraphOne [33]	In-memory NVMe SSD	Custom	Streaming, can't get arbitrary historic views if transaction time is assumed
Auxo [17]	Main and External Memory	Custom	Transaction time
[5]	Main Memory	Custom	Transaction time, Snapshot-based, focus on space savings
[2]	Main Memory	Neo4j	Valid time, in addition to entity evolution it supports schema evolution
TGraph [21]	Main and External Memory	Neo4j	Support ACID Transactions, slow topological updates but fast property updates, Transaction time
VersionTraveller [24]	Main Memory	based on PowerGraph static graph management system	Offline Snapshot-based, focus on switching between snapshots
NVGraph [39]	Non-Volatile Main Memory and DRAM	Custom	Online Snapshot-based, Transaction time
GraphVault [4]	Main and External Memory	LMDB (Lightning Memory-Mapped Database)	Uses time intervals with transaction time
AeonG [20]	Main and External Memory	Custom	Time intervals, Transaction time
Aion [61]	Main and External Memory	Neo4J + Custom Indexes	Time intervals, Transaction time

Table 2. Distributed systems for historical graph management. The second column ("Storage Model") shows whether they use a custom-based model or an already implemented one (like Neo4j). The third column ("Time-related characteristics") shows the assumptions regarding the notion of time.

Summarizing the Characteristics of Distributed Temporal Graph Management Systems		
Systems	Storage Model	Time-related characteristics
Portal [46]	Spark's Dataframes	Offline, time as a property, valid time
GDBAlive [41]	Cassandra	Transaction time
Graphsurge [53]	Custom	Offline snapshots, focus on differential computation across multiple snapshots
TEGRA [23]	Custom	Transaction time, based on persistent trees, incremental computation model, window analytics
GraphTau [22]	Spark's Dataframes	Streaming
Immortalgraph [42]	Custom	Transaction time, snapshot-based, focus on locality-aware (w.r.t. time and topology by replication), batch scheduling for computation
HGS [29]	Cassandra	Transaction time, sophisticated based on snapshot
SystemG-MV [62]	IBMs SystemG	Relaxed transaction time
Raphtory [60]	Custom + Cassandra for archiving	Transaction time, streaming
Chronograph [7]	MongoDB	Offline, time as a property, focus on graph traversals
Graphite [14]	Apache Giraph	Offline, application to time-dependent and time-independent algorithms
Granite [49]	Based on Graphite	Focus on temporal path queries, partition techniques to keep everything in main memory
Tink [38]	Apache Flink	Online, valid time
Gradoop - TPGM [9,51,52]	Apache HBase/ Accumulo	Valid and transaction time (bitemporal), fully-fledged system ranging from a graph analytical language to the storage model
Greycat [19]	NoSQL Database + custom	Valid time, no edge attributes
PAST [11]	based on key/value stores (e.g., Cassandra)	Streaming spatio-temporal graphs, bipartite graphs, only edges with time-points, spatiotemporal-specific query workloads
HINODE [31,32,58]	Custom (other versions are based on Cassandra and MongoDB)	Online, time as a property, valid time (allows more general notions of time), pure vertex-centric storage model

follows the SQL:2011 standard. They also discuss possible algorithms on temporal graphs among which are node influence over time, graph centrality over time, communities over time, and spread of information. TGraph is a valid time model that extends the property graph model (each edge and vertex is associated with a period of validity), while all relations must meet 5 criteria: uniqueness of vertices/edges, referential integrity, coalesced, required property, and constant edge association. TGA is both snapshot and extended snapshot reducible presenting a new primitive (resolve) while containing operators like trim, map, and aggregation. Portal uses Spark for in-memory representation and processing while it uses Apache Parquet for on-disk data layout using node files and edge files (but it doesn't support an index mechanism). They experimented with different in-memory representations: a) SnapshotGraph (SG), which stores the graph as individual snapshots, b) MultiGraph (MG), which stores one single graph by storing one vertex for all periods and one edge for every time period and c) OneGraph that stores each edge and vertex only once. It has distributed locality like Immortalgraph [42], and they experimented with different partitioning methods (equi-depth partitioning is more efficient in most experiments). They store materialized nodes/edges instead of deltas and they also experimented with both structural and temporal locality, concluding that temporal locality is more efficient, among other reasons, due to the lack of sufficient discrimination in the temporal ranges of the datasets.

ImmortalGraph. [42] is a parallel in-memory storage and computation system for multicore machines in a distributed setting designed for historical graphs. It focuses on locality optimizations, both in the storage of the data and in the execution of the queries using locality-aware batch scheduling (LABS). It supports parallel temporal graph mining using iterative computations. ImmortalGraph supports both global and local queries at a point in time or a time window. Data are stored in snapshot groups using either edge files or vertex files, depending on the application. A snapshot group organizes together snapshots of a time interval by storing the first snapshot and then logging the rest of the changes. Either time locality can be ensured by grouping activities associated with a vertex (and a vertex index) or structural locality can be ensured by storing together neighboring vertices (and a time index). In order to combine the advantages of both approaches, they replicate the needed data and decide which approach to use according to the type of query and the distance from the snapshot of the group. LABS favors partition-parallelism from snapshot-parallelism, so they prefer batch operations of vertex/edges achieving better locality and less inter-core communication. They also experimented with iterative graph mining and iterative computations. For the former, they reconstruct the needed snapshots in memory favoring time locality (and they compare both push, pull, and stream techniques), while for the latter they compute the first snapshot and the succeeding snapshots in batch (achieving better locality). They also implemented both low-level and high-level query interfaces, the latter used for iterative computations. An earlier implementation of ImmortalGraph is Chronos [18] with the main difference being that it only focuses on time locality. Finally, they provide a low-level as well as a high-level programming interface (APIs), which can be considered as an analytics engine. They experiment on Pagerank, diameter, SSSP, connected components, maximal independent sets, and sparse-matrix vector multiplication.

Historical Graph Store (HGS). [29] is a cloud parallel node-centric distributed system for managing and analyzing historical graphs. HGS consists of two major components, the Temporal Graph Index (TGI), which manages the storage of the graph in a distributed Cassandra environment, and the Temporal Graph Analysis Framework (TAF), which is a Spark-based library for analyzing the graph in a cluster environment. TGI combines partitioned eventlists, which store atomic changes, with derived partitioned snapshots, which is a tree structure where each parent is the intersection of children deltas (used for better structure locality storing neighborhoods). Both of them are partitioned, while they are also combined with a version chain to maintain pointers to all references of nodes in chronological order. TGI divides the graph into time spans (like the snapshot groups of ImmortalGraph) with micro-deltas, which are stored as key-value pairs contiguously into horizontal partitions at every time span. In this way, it can execute in parallel every query in many query processors and aggregate the result to the query manager or to the client. It can work both on hash-based and locality-aware partitioning by projecting a time range (time-span) of the graph in a static graph. TAF supports both point-in-time queries and time-window queries. Some of the supported queries are subgraph retrieval with filtering, aggregation, pattern matching, and queries about the evolution of the graph. An earlier implementation of TGI is DeltaGraph [27], which focuses on snapshot retrieval.

ChronoGraph. [7] is a temporal property graph database built by extending Tinkerpop and its graph traversal language Gremlin so as to support temporal queries. It stores the temporal graph in persistent storage (MongoDB), and then loads the graph in-memory and traverses it. Their innovation is not in the storage model but in how they support traversal queries efficiently on top of it. It exploits parallelism, the temporal support of Tinkerpop to increase efficiency, and lazy evaluations to reduce memory footprints of traversals. Its main focus is on temporal graph traversals but can also return snapshots of the graph. They distinguish point-based events and period-based events because of their semantics and their architectural needs. They use aggregation to convert point-based events to period-based events so as not to have two different semantics in order to improve time efficiency in query execution. They achieve this by using a threshold as the maximum time interval that may exist between time instants so as to group them together. A graph is composed of a static graph, a time-instant property graph, and a time-period property graph. They also use event logic, where an event might be either a vertex or an edge, during a period of time or a time instant. They implemented temporal versions of BFS, SSSP, and DFS. However, they don't recommend DFS on their system because of Gremlin's recursive logic.

An extension of Chronograph by using time-centric computation for traversals is given in [6].

Tink. [38] is an open-source parallel distributed temporal graph analytics library built on top of the Dataset API of Apache Flink and uses the programming language Gelly. It extends the temporal property graph-model focusing on keeping intervals instead of time-points by saving nodes as tuples. It depends on Flink to use parallelism, optimizations, fault tolerance, and lazy-loading and supports iterative processing. It also uses functions from Flink like filtering, mapping, joining, and grouping. Most algorithms use Gelly's Signal/Collect (scatter-gather) model, which executes computations in a vertex-centric way. It also provides temporal analytic metrics and algorithms. For the latter, they implemented shortest path earliest arrival time and shortest path fastest path while for temporal metrics they provide temporal betweenness and temporal closeness.

Gradoop (TPGM). Temporal Property Graph Model (TPGM) [9,51,52] is an extension of Gradoop's Extended Property Graph Model - EPGM - (model for static graph processing, presented in a series of papers from 2015, see [25]) to support temporal analytics on evolving property graphs (or collection of graphs) that can be used through Java API or with KNIME. Gradoop is an open-source parallel distributed dataflow framework that runs on shared-nothing clusters and uses GRALA as a declarative analytical language and TemporalGDL as a query language. Gradoop supports Apache HBase, and Apache Accumulo to provide storage capabilities on top of HDFS, while other databases can also be used with some extra work. TPGM supports bitemporal time by maintaining logical attributes for start and end time for both valid and transaction time for every graph entity. While TPGM provides an abstraction, Apache Flink is used for handling the execution process in a lazy way. GRALA provides operators both for single graphs and graph collections, it supports retrieval of snapshots, transformations of attributes or properties, subgraph extraction, the difference of two snapshots, time-dependent graph grouping, temporal pattern matching, and others. For some more complex algorithms, it also supports iterative execution using Apache Flink's Gelly library. Lastly, they have implemented in Flink a set of operations for their analytics engine - by using Flink Gelly. They provide an extensive description of their architecture, while they also provide a *Lessons Learned* section that contains valuable information concerning their design choices.

SystemG-MV. In [62] they propose an OLTP-oriented distributed temporal property graph database (dynamically evolving temporal graphs). It is built on top of IBM's SystemG, which is a distributed graph database using LMDB (*B*-tree-based key-value store). Data are stored in tables with key/value pairs allowing to query part of the graph efficiently without retrieving the whole snapshots. Different tables exist for vertices, edges, and properties, while it supports updates only on present/future timestamps like transaction-time models. Therefore, changing previous values of the graph is not allowed explicitly, but it is possible to change past events by using low-level methods. In this model, they save two timestamps for the creation/deletion of vertices/edges but they don't allow edges to be recreated with the same id, although multiple edges can exist between a pair of vertices. For vertices, they keep the deleted vertices in a different table, while for properties they keep only one timestamp per update, as the rest can be calculated. Alongside the historic tables, they keep one table with the current state of the graph for more efficient queries. Their low-level algorithms that constitute the API of their storage model resemble the API suggested in [32].

TEGRA [23] is a distributed system with a compact in-memory representation (using their custom storage model) both for graph and intermediate computation state. Its main focus is on time window analytics for historical graphs, but it can also be used for live analytics as the data are ingested in the database. An interesting feature is the ICE computational model that takes advantage of the intermediate state of computations by storing it and using it in the same or similar queries. Computations are performed only on subgraphs affected by updates at each iteration. This has some overhead related to finding the correct state and the extra entities that should be included in the query when there are many updates at each iteration or while trying to use ICE on different queries. Tegra also uses TimeLapse, an API for high-level abstraction that allows what-if questions that change the graph creating different histories, suited for data analytics purposes. The storage model behind TEGRA is DGSI, which uses persistent data structures to maintain previous versions of data during updates. In particular, it uses persistent Adaptive Radix Trees (pART) to store edges and nodes separately by employing the path-copying technique [12]. It uses simple partitioning strategies to distribute the graph to nodes. Log files are used to store updates between snapshots, which are stored in turn in the two pARTs. They use branch-and-commit primitives in tandem with the GAS (Gather - Apply - Scatter) model [16]. It also supports changing any version, thus leading to a branched history like a tree (reminiscent of full persistence [12]). Lastly, TEGRA also uses an LRU policy to periodically remove versions that have not been accessed for a long time.

Graphite. [14] is a distributed system for managing historical graphs (offline with valid time) by using an interval-centric computing model (ICM) built over Apache Giraph. They assume data are given in ascending time order and any vertex can exist only once for a contiguous time interval. It can execute both time-independent and time-dependent historical queries (temporal queries on a time window). They introduce a unique time-warp operator for temporal partitioning and grouping of messages that hides the complexity of designing temporal algorithms while avoiding redundancy in user logic calls and messages sent. ICM uses Bulk Synchronous Parallel (BSP) execution for every active vertex of a query until it converges. They use two stages of logic, compute and scatter, where compute does the computations needed for a vertex, and scatter transfers it with messages to neighbor vertices as needed. The time-warp operator is applied at the alternating compute scatter steps to help the sharing of calls and messages across intervals. A key aspect is that it correctly groups the input with no duplicates, while it returns the minimum possible number of triples. They also designed and constructed a plethora of time-independent and time-dependent algorithms for their system with an extensive experimental evaluation.

Granite. [49] is a distributed engine for storing and analyzing temporal property graphs (supports temporal path queries) made on top of and as a sequel to Graphite focusing on path queries. Its design is based on the assumption that the workloads consist of infrequent updates and frequent queries. They extend the previous model by adding a temporal aggregation operator, indexing, query planning, and optimization, while they prefer to relax ICM so as to make it work beyond time-respecting algorithms. Granite handles both static temporal graphs and dynamic temporal graphs while it uses interval-centric features only in the latter. To optimize path queries, they split them and execute them concurrently, while they also keep statistics about the active nodes at each time point so as to optimize the query planning. While Graphite makes hash partitioning at query execution, Granite first partitions every entity according to its type, and then it performs a topological partition to its independent group of entities of the same type and splits them into workers using the round-robin technique. They also use a result tree so as not to send duplicate paths across the system (some parts of the path might be the same). Lastly, they propose a query language for path queries.

3 Preliminaries

Let $G = (V_T, E_T)$ be a static historical network. The set of historical nodes V_T consists of a set of nodes along with their time intervals, that is $V_T \subset V \times \mathbb{N}^2$. The set of historical edges E_T is a set of edges along with their time intervals, that is $E_T \subset E \times \mathbb{N}^2$, where E contains all possible $\binom{|V|}{2}$ undirected edges. Note that we consider nodes and edges that have a single valid time interval, but it is easy to generalize to a set of valid time intervals. The state of the graph G at a particular time instant t, is called a snapshot and it is denoted by G_t. The preceding definitions mean that each node $v \in V$ (and edge $e \in E$) has a time interval attached $[t_v^{(s)}, t_v^{(f)}]$ (similarly $[t_e^{(s)}, t_e^{(f)}]$) (where (s) and (f) stand for start and finish respectively) that dictates the time instants where node v (edge e) is existent. Thus, if $t \notin [t_v^{(s)}, t_v^{(f)}]$, then v is not existent at time t. $V_{ij} \subseteq V$ contains all nodes that have a time interval that spans the query interval $[t_i, t_j]$. The time interval of each edge is by definition a subset of the time interval of the respective vertices. In case of multiple time intervals, we have to define the borders of each interval accordingly, to avoid overlaps. For example, each interval should be open at the left and closed at the right. The convention we make is that a time point t is represented by $(t, t]$. Assume that by $\mathcal{N}_{ij}(v)$ we represent the neighborhood of node v in the query time interval $[t_i, t_j]$. Note that $\mathcal{N}_{ij}(v)$ may even be the empty set for specific query time intervals.

3.1 HiNode Theoretical Model

In [32], the first purely entity-centric, and more specifically, vertex-centric model for maintaining graph historical data, termed HiNode is introduced. Its strongest point is that it builds upon a theoretical storage model that is asymptotically space-optimal. The core idea behind HiNode's solution is that a vertex history throughout all snapshots is combined into a set of collections called diachronic node. The diachronic node utilizes indexes to model a vertex's history as a collection of intervals that permit geometric operations upon them. In particular, a vertex $v \in G_i$ is characterized by a set of attributes (e.g., color), a set of incoming edges from the other vertices of G_i, and a set of outgoing edges to the other vertices of G_i. We construct an external interval tree $\mathcal{I}$ that maintains a set of intervals $[t_v^{(s)}, t_v^{(f)}]$ for each vertex v. We mark a vertex to be "active" (alive) up until the latest time instant, by setting the $t^{(f)}$ value to be $+\infty$. Finally, each interval $[t_v^{(s)}, t_v^{(f)}]$ is augmented with a pointer (handle) to an additional data structure for each vertex v, corresponding to the diachronic node.

A diachronic node $\mathcal{D}_v$ of a vertex v maintains a collection of data structures corresponding to the full vertex history of G, i.e., when that vertex was inserted, all corresponding changes to its edges or attributes and finally its deletion time (if applicable). More formally, a diachronic node $\mathcal{D}_v$ maintains an external interval tree $\mathcal{I}_v$ that stores information regarding all of v's characteristics (attributes and edges) throughout the entire history. An interval in $\mathcal{I}_v$ is stored as a quadruple $(f, \{\ell_1, \ell_2, \ldots\}, t^{(s)}, t^{(f)})$, where f is the identifier of the attribute that has values $\ell_1, \ell_2, \ldots$ during the time interval $[t^{(s)}, t^{(f)}]$. Note that an edge of v (i.e., one endpoint of the edge is v), can be represented as an attribute of v by using one value ℓ_i to denote the other end of the edge, another value ℓ_j to mark the edge as incoming or outgoing and a last value ℓ_h that is used as a handle to the diachronic node corresponding to the vertex in the other end of the edge. The remaining ℓ values can be used to store the attributes of the edge themselves (e.g., weight). Additionally, the diachronic node maintains a B-tree for each attribute and for each individual edge of the vertex. Full details are in [32].

HiNode supports adding or removing vertices and attributes as fundamental operations upon which more complex operations and queries (e.g., graph traversal, shortest path evaluation, etc.) are constructed. In HiNode, each change is stored $O(1)$ times, resulting in an asymptotically optimal total space cost. Furthermore, due to the local handling of history, HiNode performs well on local queries and the authors further demonstrate that HiNode on top of G^* is competitive regarding global queries as well compared to G^* [32].

3.2 Implementation in Cassandra

The first HiNode implementation, hereafter termed as HiNode-G^*[2], was based on extensions to the G^* [34,57] distributed graph database. This design choice

---

[2] Source code available at https://github.com/hinodeauthors/hinode.

incurred severe limitations regarding the efficiency and scalability of the HiNode-G^* prototype. In a follow-up work [31], to outperform solutions based on tailored graph management systems, such as Neo4j, we proposed to leverage NoSQL as the underlying database technology providing preliminary results with respect to two different implementation approaches in Cassandra. These approaches consist of the Single Table (ST) and Multiple Table (MT) implementations. In the former case, the entire history of a vertex is stored in a single table with each vertex corresponding to a single table row, while in the latter case, the data of each vertex is stored in multiple tables with each table corresponding to a single vertex attribute. [3]

To adequately support global queries (i.e., queries that involve a significant part of a snapshot's vertices), the two models offer two querying modes for the retrieval of all vertices relevant to a specified query. Let $[t_s, t_e]$ be a specified time range for which a query is about to be executed. In the first mode (termed Retrieve_All (RA)), and regardless of the given time range, we retrieve all vertices from each model and then perform a client-side filtering operation, where we discard any vertices that do not belong in $[t_s, t_e]$. In the second mode (termed Retrieve_Relevant (RR)), in each model, we first determine the vertices that are "alive" at $[t_s, t_e]$, and then, we retrieve them.

While in ST the implementation of RR is straightforward, MT requires additional work since retrieving a particular (set of) attribute(s) during a certain time interval $[t_s, t_e]$ would translate to a range query and the retrieval of all data with a "timestamp" value between t_s and t_e (i.e., we are not interested in any updates that occur outside $[t_s, t_e]$). Since Cassandra does not natively permit double-bounded range queries for the sake of efficiency, we fetch the relevant data with a timestamp larger than t_s and then filter all data with a timestamp larger than t_e at the client side. In [31] there is extensive experimental evaluation, with interesting outcomes which are not presented here due to space limitations. All in all, the choice of a particular vertex-centric implementation is not straightforward and exhibits different trade-offs depending on the query at hand.

4 The MongoDB Implementation

Our main motivation behind using MongoDB is to exploit the wider range of indexing options and its capabilities to reduce client involvement when processing queries. Additionally, in Cassandra, data are saved as strings and, as such, they are serialized when returned to the client, while in MongoDB we can store the elements of the nodes with a combination of lists and documents. Overall, we can perform more complex in-database queries and decrease client involvement in query processing. Finally, in the new implementation, instead of getting the documents from the database in a single large batch, we have the option to

---

[3] Source code available at https://github.com/akosmato/HinodeNoSQL.

employ a `foreach` approach (when this is expected to be more efficient) and as a result, to mitigate intermediate client-side storage requirements[4].

4.1 Schema Alternatives

Both the ST and MT models have been transformed to comply with MongoDB's JSON format in a straightforward manner. In addition, we developed an alternative schema for both models, where the elements of the primary key are inserted as characteristics in the document; the primary key is the standard key assigned automatically by MongoDB. The reason for this schema is to further simplify the client-side tasks (i.e., the processing refers to the document content exclusively) with no difference in the capability of answering specific types of queries.

In the ST model, a document is a representation of a diachronic node and consists of the primary key as a triple (`vid, start` and `end` of the node), the incoming and outgoing edges, and the vertex attributes. The features forming the key are stored as atomic string values, while the vertex attributes are stored as a list of sub-documents, where each document is a triple. The incoming and outgoing edge metadata are stored as a sub-document containing a list of triples (where each triple is a MongoDB sub-document). The former document is essentially a hashmap structure with the key corresponding to the vertex id, while the nested sub-document stores the attributes and the period for each edge. The following three indices are built: (i) an index on `vid`; (ii) an index on `start` and `end`; (iii) an index on the complete key. The first index allows quick retrieval of a specific vertex, while the second and third indices facilitate stabbing queries.

In the MT model, we split the diachronic node into three sets of collections, one containing vertices, one containing incoming edges, and one containing outgoing edges. Each set consists of one collection concerning the existence period of the vertex or edge and one collection for every attribute. The standard indices are on `vid`, (`vid,start`) and (`vid,start,end`) in the first set of collections. For the edge collection sets, the multikey indices are on `sourceID` (or `targetID`) and the `start` timestamp.

In summary, the main difference with the Cassandra-based implementation in [31] in terms of modelling is the increased flexibility regarding indices and the fact that sub-documents are stored without being serialized as strings, thus making able to run more complex queries.

4.2 Query Processing

For local queries, the server (database) side is straightforward, while most of the work is performed on the client side. The local queries we investigate in this paper are the retrieval of the history of a vertex and one-hop queries. In the former query, we must retrieve the history of a specified vertex for a time interval. In the latter query, we must retrieve the neighbors of a vertex at a

---

[4] Source code available at https://github.com/alexspitalas/HiNode-MongoDB/.

specified time interval. Both queries are straightforwardly supported by both implementation models.

Because of the vertex-centric approach, global queries are more meaningful to investigate in depth, aiming at rendering them more efficient. Global query processing is comprised of two phases. The first phase is related to the retrieval of the data, while the second phase is related to the processing of the retrieved data. These phases can be intertwined. In our implementation, the two phases are separated so that the client's side is the same for all ST-based and all MT-based approaches, respectively. Regarding the retrieval of the data, three variants have been developed, Retrieve_Relevant (RR), Retrieve_All (RA), and In-Database (ID).

In RR, the main objective is to find the relevant documents by retrieving only their necessary characteristics. In the RA approach, we retrieve all the characteristics of the document, while we check if the document is needed for the query. In ST model, a document consists of all information of a diachronic node (edges and properties), while in MT model, a document is referring to either a node, an edge or a property and the corresponding information in each case. Comparing this method with RR, we perform only one read at the database, but we retrieve more data than necessary if the document is not needed for the query; as a result, RR is expected to perform better when the amount of data stored per node is much higher than those needed to establish if the node is relevant to the query. This relevance check, along with the rest of the query execution, is performed on the client. In the new MongoDB implementation, contrary to the initial implementation based on Cassandra, we adopt a more incremental (iterative) approach instead of returning all data in a single batch; this has increased the scalability of global queries so that they can be executed without throwing an out-of-memory error.

However, the most notable difference between the two implementations is that MongoDB naturally lends itself to in-database query processing, so that the client gets only the data needed to compute the final results and not a superset of these data (through submitting more complex queries as supported by the MongoDB driver). To this end, we use the in-database MongoDB mechanisms to perform the relevance checks mentioned in the RR approach. Similarly to RR, the data needed for the final answer computations are returned incrementally to the client. As such, this approach has even lower space requirements on the client side, and at the same time, it allows for both the server and the client side to work in parallel[5].

4.3 Transactions with MongoDB

In a distributed environment, supporting both OLTP and OLAP queries in the same database is challenging, although some efforts have been made in this area.

---

[5] In some local queries (like one-hop query), it may make sense to adopt an in-database query processing rationale, but this is beyond the scope of this paper.

In RDBMS databases, OLTP queries are supported by ensuring ACID properties. However, these properties are more difficult to maintain in a distributed environment, especially in NoSQL databases. In our case, the specifics of the OLTP and OLAP queries we aim to support are summarized below.

- Support of OLTP queries: These are simple queries typically involving a few records. The emphasis is on fast processing, as OLTP databases are frequently read, written, and updated. If a transaction fails, the built-in system logic ensures data integrity.
- Support of OLAP queries: These are complex queries usually referring to a large part of the graph. The emphasis here is on efficiency in executing the queries without concern for parallel updates performed on the graph.

To address these requirements, MongoDB supports multi-document ACID transactions with minimal additional code. Due to multi-document ACID support, if a transaction involves multiple documents, it will maintain Atomicity, Consistency, Isolation, and Durability throughout the operation, leaving the database in a consistent state. This feature allows us to support transactions in our MT model with MongoDB. To the best of our knowledge, some NoSQL databases only support ACID properties up to one document. In contrast, many NoSQL databases do not support ACID transactions at all, instead adopting a BASE transaction model that prioritizes availability over consistency.

5 Experiments

Experiments were carried out both locally and in a cluster, with the aim of obtaining different conclusions in each setting. The single node system has an AMD Ryzen 9 7900X CPU @ 4.70 GHz, 196 GB DDR5 RAM, and a 1 TB NVMe, while the client and the databases are co-located on the same machine. The client application was written in Java.

For the cluster experiments, we used a 4-node cluster, where all the machines of the cluster have the following specifications: Intel(R) Core(TM) i7-10700 CPU @ 2.90 GHz, 16 GB RAM, and a 1 TB NVMe disk. The client and the database are co-located on the same network with a 10 GBps LAN connection but in a different machine. The client machine is equipped with an Intel(R) Core(TM) i3-7100 CPU @ 3.90 GHz, 4 GB RAM, and a 500GB HDD.

For the experiments, we employ two queries, one local and one global query: 1) *one-hop* (local query): that returns all nodes that are adjacent to a given node within a time interval, 2) *vertex degree distribution* (global query): that returns the degree distribution of a graph within a time interval. These queries are applied on three different datasets that are shown in Table 3. We experimented with all different combinations of the MT and ST models, and the Cassandra and MongoDB systems. For MongoDB, we test with all modes of global query processing (retrieve-all, retrieve-relevant, in-database). Each query refers to a range of snapshots from 1 to *all*.

Table 3. The datasets used in the snapshot experiments.

Name	# of vertices	# of edges	# of snapshots
hep-Th [36]	27.77K	352.8K	156
hep-Ph [35]	34.5K	421.6K	132
US Patents [37]	3774.8K	16.5M	444

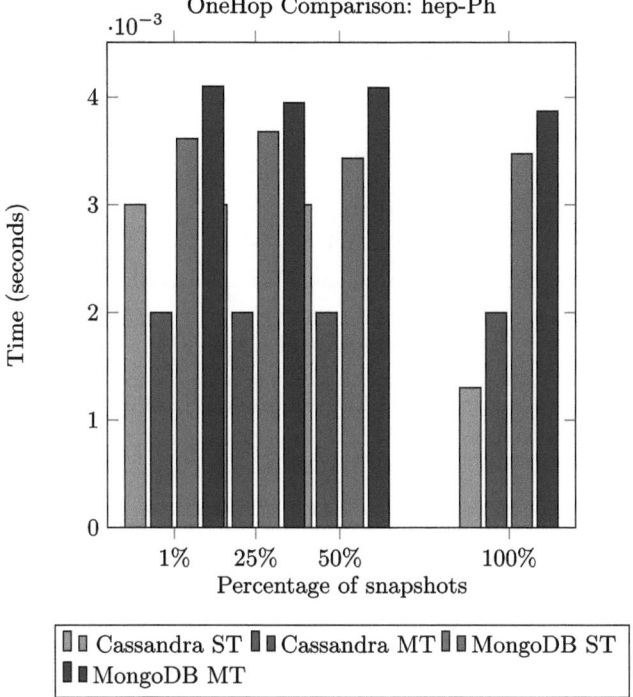

Fig. 1. OneHop query comparison for the hep-Ph dataset in the Cluster.

5.1 Experiments on Snapshot-Based Datasets

For the snapshot-based datasets, the temporal domain is discretized into a finite set of discrete time instants within which, all time-relative entities are aligned. This approach offers increased query efficiency and facilitates conversion of the dynamic graph into a static one at a specific time-instant. However, it is not always feasible to represent the problem using snapshot-based datasets due to certain limitations.

Local Queries. Regarding local queries on a cluster environment, we executed one-hop queries for all three snapshot-based datasets, retrieving data from 1 snapshot, 25% of the snapshots, 50% of the snapshots and 100%, so as to observe

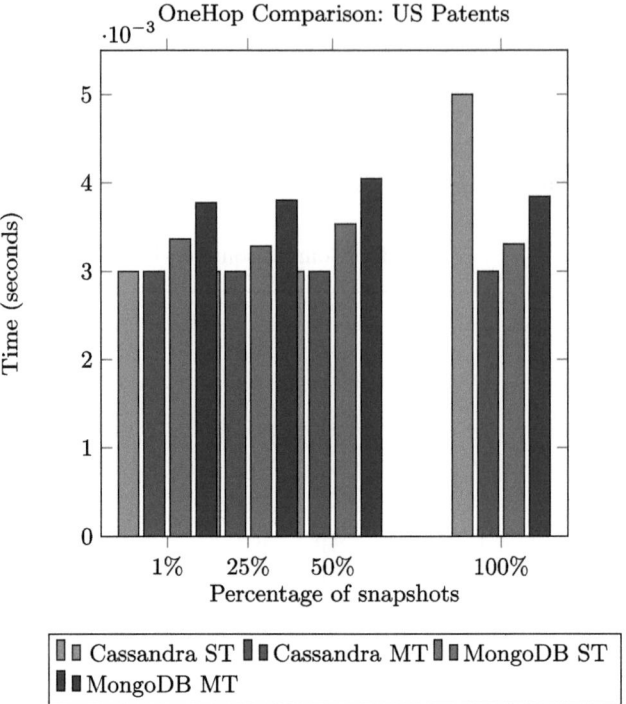

Fig. 2. OneHop query comparison for the US Patents dataset in the Cluster.

the scaling of the models while we retrieve more historical data. We repeated each query 500 times and we report the average values. For each set of 500 runs we have a cold start. In Figs. 1, and 2 we observe the performance of every model in each dataset, reaching the following conclusions (Hep-Th has similar results with Hep-Ph and it is not depicted):

1. The most important observation is that in general, Cassandra MT outperforms Cassandra ST, managing to decrease time up to 84.6%, (the gap decreases as we move to bigger datasets). However, MongoDB ST slightly outperforms MongoDB MT, managing to decrease time by 6.7% up to 24.9%. This difference can be attributed to the fact that MongoDB is a document store, and as such, it can better cope with complex collections like in the case of ST, where all data of a node are placed together.
2. Another observation is that in MongoDB, the performance is stable as we increase the percentage of snapshots used in the query. This is expected in a local query at a vertex-centric system, while in Cassandra there are some spikes when we use 100% of the snapshots.
3. While querying for up to 50% of the snapshots, Cassandra MT is the best overall model, managing to decrease the running time between 8.6% up to 45.6%. However, when the query involves 100% of the snapshots, the best

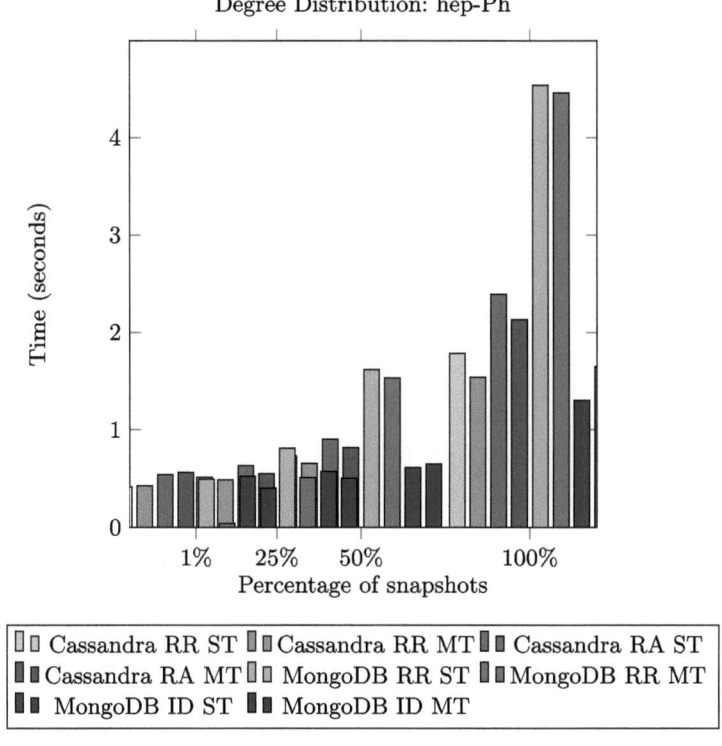

Fig. 3. Degree Distribution Comparison for the Hep-Ph dataset.

overall model is MongoDB ST, managing to reduce time by more than 50% compared to the worst model.

Global Queries. Regarding global queries, we demonstrate the results for both hep-Th, hep-Ph and US Patents datasets. Our experiments include more query processing modes than local queries, since we distinguish between RA, RR and ID approaches. Recall that we focus on global queries in this paper, since this is the expected bottleneck for a vertex-centric approach.

Vertex Degree Distribution Query. The summary results are shown in Figs. 3, 4, and 5, from which we draw the following observations:

1. Due to memory restrictions, among the Cassandra models only Cassandra ST RR was able to execute for 100% of the snapshots in the US-Patents dataset. On the other hand, all MongoDB models managed to execute for all query percentages.
2. The best models differ based on the size of the dataset and the percentage of snapshots used. Cassandra MT RR is the best Cassandra model for smaller datasets in all snapshots, while for the bigger Dataset (US-Patents) until 50% of the snapshot Cassandra MT RA is the best Cassandra model, while for

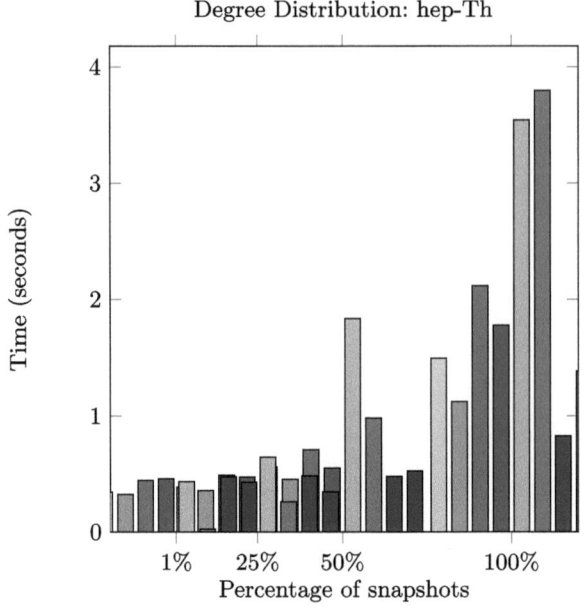

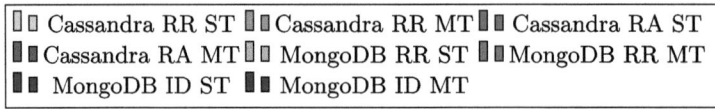

Fig. 4. Degree Distribution Comparison for the Hep-Th dataset.

100% of the snapshots Cassandra MT RR is the best Cassandra model. For MongoDB, when 100% of the snapshots are queried, the best model is MongoDB ST ID. In all cases, the best model is a MongoDB model, outperforming the best Cassandra model with a percentage difference between 3.17–31.05% in the smaller datasets, and 20.56–52.42% in US Patents.
3. In most cases, the Cassandra MT models, outperform the corresponding Cassandra ST models, with a percentage difference up to 28.31% in the smaller datasets, increasing to 102.14% in US-Patents.
4. In MongoDB in most cases the MT model outperforms the corresponding ST model, by up to 85% percentage difference, although when the query includes more than 50% of the snapshots MongoDB ST ID outperforms the corresponding MT model by up to 76.82% percentage difference.

Finally, while Cassandra requires less space to store the data since it builds fewer indexes and adopts a different storage approach, the MongoDB approach requires less memory on the client while executing the query. This is due to the iterative approach that was adopted in MongoDB, as well as due to the adoption of the ID query processing method. The space required for the three datasets in Cassandra ST was 31.0 MB, 37.4 MB and 1.83 GB, and for the MT was 45.7

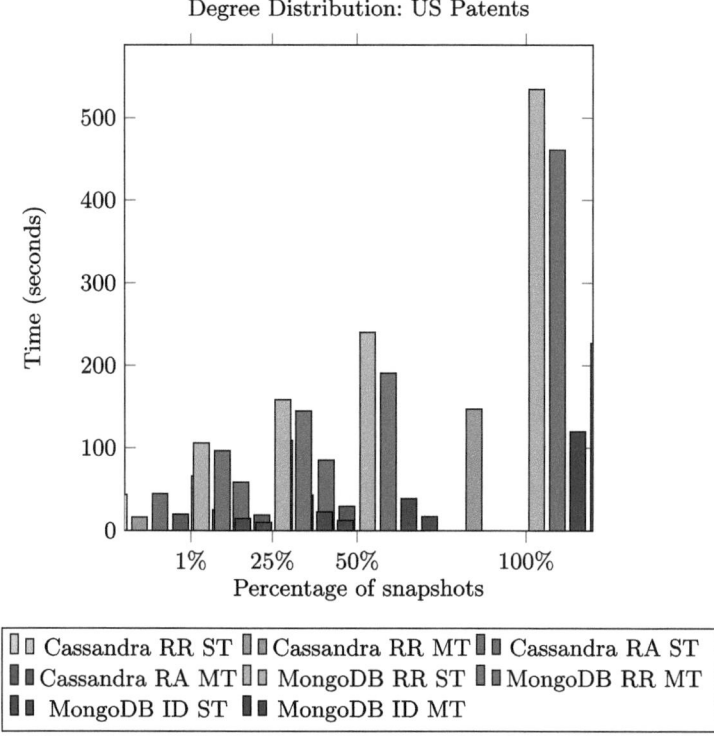

Fig. 5. Degree Distribution Comparison for the US Patents dataset.

MB, 55.5 MB and 3.10 GB, respectively. The space for MongoDB ST was 89.70 MB, 107.37 MB, and 4.84 GB, and for MT was 218.34 MB, 260.87 MB and 10.96 GB, respectively. On the other hand, MongoDB exhibited a speedup larger than 2× when inserting data. In the RR and RA techniques both in MongoDB and Cassandra, we need to store either vid or the whole documents before processing them. On the other hand, clients using the ID approach need to retrieve only the document that is currently being processed, and as a result, the storage complexity depends on the execution of the query.

Regarding global queries in the cluster environment, we demonstrate the results for the hep-Th, hep-Ph, and US Patents datasets. We employ only the Degree Distribution query, as we observed similar results with the Average Degree query. In addition, based on the previous experiments, we use only the most promising approaches. In Cassandra, we experiment with RA and RR, while in MongoDB we experiment with RR and ID both in ST and MT models.

The results can be seen in the Figs. 6, 7 and 8. In the following, we summarize the main observations from these experiments.

1. In the two smaller datasets (hep-Ph, hep-Th), the MongoDB ST ID Model is the best-performing model, managing to reduce time from 30% up to 64.6%

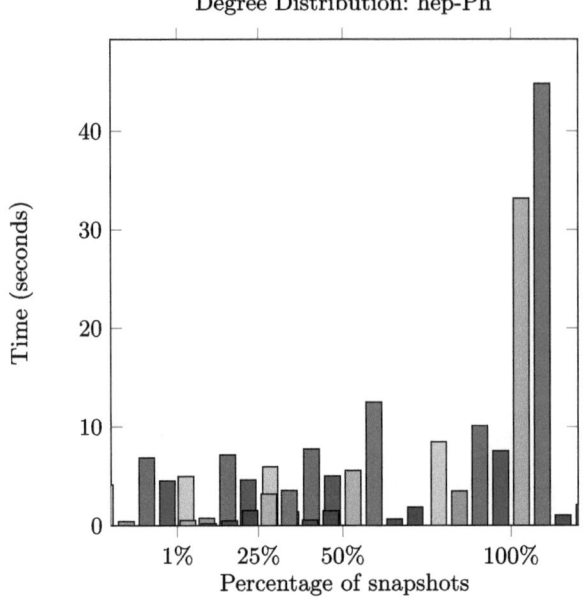

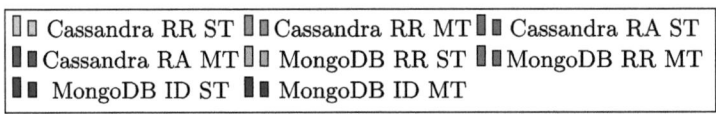

Fig. 6. (Cluster) Degree Distribution Comparison for the Hep-Ph dataset.

when compared to the best Cassandra model, with the exception when querying for only one snapshot. In this case (the query involves one snapshot), the best-performing model in the small datasets is Cassandra MT, reducing time by more than 10% when compared to the best MongoDB model.
2. In Cassandra, the MT models manage to outperform the corresponding ST models. The RR approach reduces execution time by 58.9% to 89.6%, 71.6% to 92.2% and 70% in the datasets hep-Ph, hep-Th, and US Patents respectively. Similarly, RA reduces time by 25.4% to 35.5% and 25.1% to 34.3% for the two smaller datasets.
3. In MongoDB, the ST models manage to outperform the corresponding MT models. Recall that the ST model corresponds to a pure vertex-centric approach as in the case of HiNode. The RR approach reduces execution time by 10.1% to 59.9%, 20% to 26.7%, and 10% to 29.6% in the datasets hep-Ph, hep-Th and US Patents correspondingly. Similarly, ID reduces time by at most 70.2% in hep-Ph, 62.4% to 72.9% in hep-Th, and 4% to 33.4% in US Patents.
4. Due to low memory on the client side, some experiments on the US Patents dataset failed. This indicates which models and algorithms are less

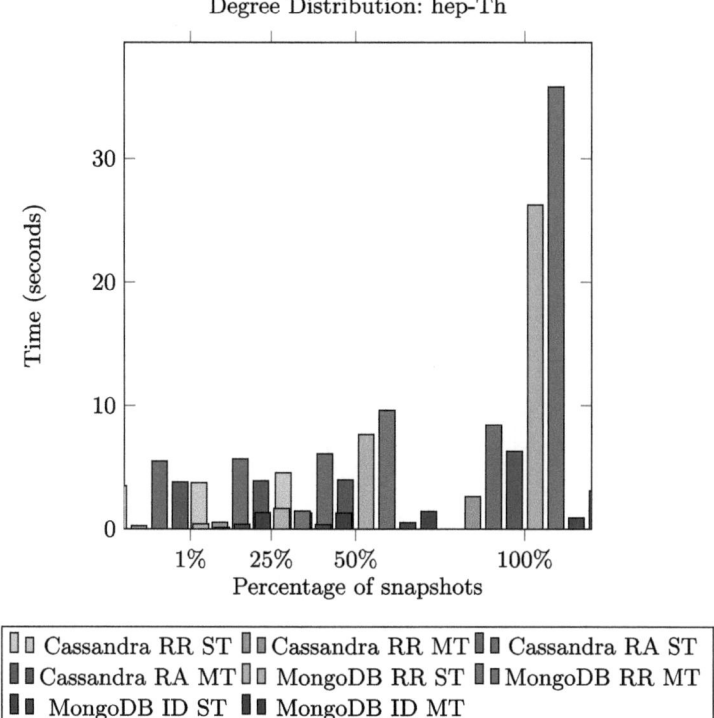

Fig. 7. (Cluster) Degree Distribution Comparison for the hep-Th dataset.

Table 4. The database creation time from a sequence of transactions (insertions, deletions, modifications) for each model and each dataset in a local machine (in minutes).

		hep-Th	hep-Ph	US Patents
MongoDB	ST	2.6	3.0	118.3
	MT	1.96	2.4	105.27
Cassandra	ST	3.91	4.44	191.2
	MT	3.72	4.44	188.2

memory-demanding for the client and increase the distribution of tasks. Those are the MongoDB RR ST, MongoDB RR MT, and MongoDB ID ST that were able to execute the query for up to 50% of the graph snapshots of the US Patents dataset.

Transactions. Regarding testing transactions, our experiments involve the operations of insertion, modification, and deletion. These actions are performed at the beginning of our tests while creating the database. No tests were made with other OLAP queries (e.g., one hop-queries) that could be executed

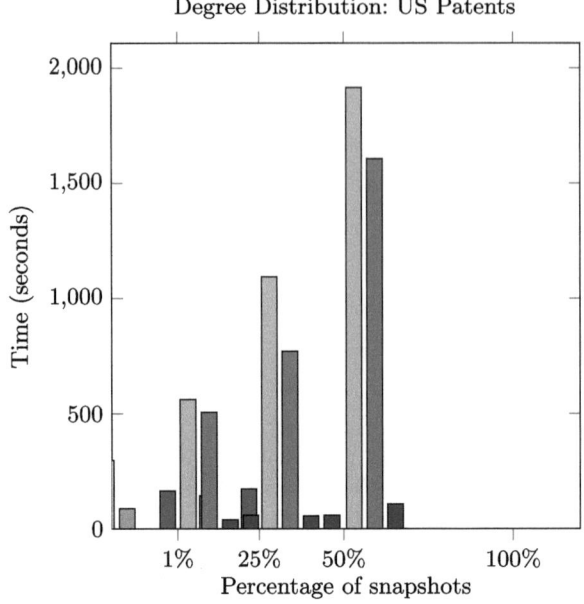

Fig. 8. (Cluster) Degree Distribution Comparison for the US Patents dataset.

Table 5. "The database creation time from a sequence of transactions (insertions, deletions, modifications) for each model and each dataset in the cluster (in minutes).

		hep-Th	hep-PH	US Patents
MongoDB	ST	25	37.9	1318.8
	MT	58.7	63.4	2879.4
Cassandra	ST	40.3	42.5	1747.5
	MT	49.8	58.7	2485.1

concurrently. In Table 4, we provide the total database construction time for each model in local mode (single machine). By observing the results we can reach the following deductions:

1. The ST models have comparable efficiency with the MT models, with less than 13% time reduction when MT model is used.
2. Comparing the best Cassandra with the best MongoDB models, the latter outperforms Cassandra by 79.1% to 89.79%.

In Table 5 we report on the construction times in the cluster environment. We observe a rather different behavior compared to inserting the data locally, which may be attributed to many factors. The most important reasons for why the ST model outperforms the MT model is the replication of data in the cluster, the persistence of data throughout the cluster, as well as the use of the majority consistency level. In addition, since we used strict rules in the cluster in order to better support the consistency of the data, it is expected that we will have slower transaction times.

1. The ST models outperforms the MT models, by 118.43–134% in MongoDB and 23.8–42.35% in Cassandra.
2. Comparing the best Cassandra with the best MongoDB models, the latter outperforms Cassandra by 32.54–61.2%.

5.2 Experiments on a Historical Graph Dataset with Time Intervals

To enhance the support for our experimental investigations, we sought to evaluate our models utilizing workloads designed for historical graph analysis. Upon an exhaustive examination of available resources in this domain, it became evident that there is a notable scarcity of openly accessible datasets containing historical information conducive to our research objectives. Consequently, we tried to come up with alternative solutions.

The Linked Data Benchmark Council (LDBC) [3], renowned for its contributions to benchmarking workloads and dataset generation, has developed a suite of tools tailored for graph database evaluation. Among these tools, the "LDBC SNB Datagen" generator stands out for its provision of comprehensive information regarding data creation and deletion events. Although this dataset does not encompass updates on parameters, it aligns closely with our current dataset test case and enjoys widespread recognition within the research community.

While historical information pertinent to our research goals is absent from the interactive LDBC SNB workload, it is available within the dataset generated by the aforementioned data generator. To leverage this dataset as a testing workload, we needed to move to an event-based paradigm. The original format of the data, wherein all node and edge information, including parameter data and temporal details, are encapsulated within a single row, posed a challenge. To better emulate real-time operational scenarios, we restructured the dataset and ordered the events chronologically, primarily impacting the end time[6].

The Schema of the Historical Graph Dataset. The dataset generated from the Linked Data Benchmark Council (LDBC) for Social Network Benchmark (SNB) exhibits diverse forms contingent upon user configurations. Users possess the capability to adjust cardinality numbers, as well as the minimum and

---

[6] This transformation process is publicly available in https://github.com/alexspitalas/HGDataset.

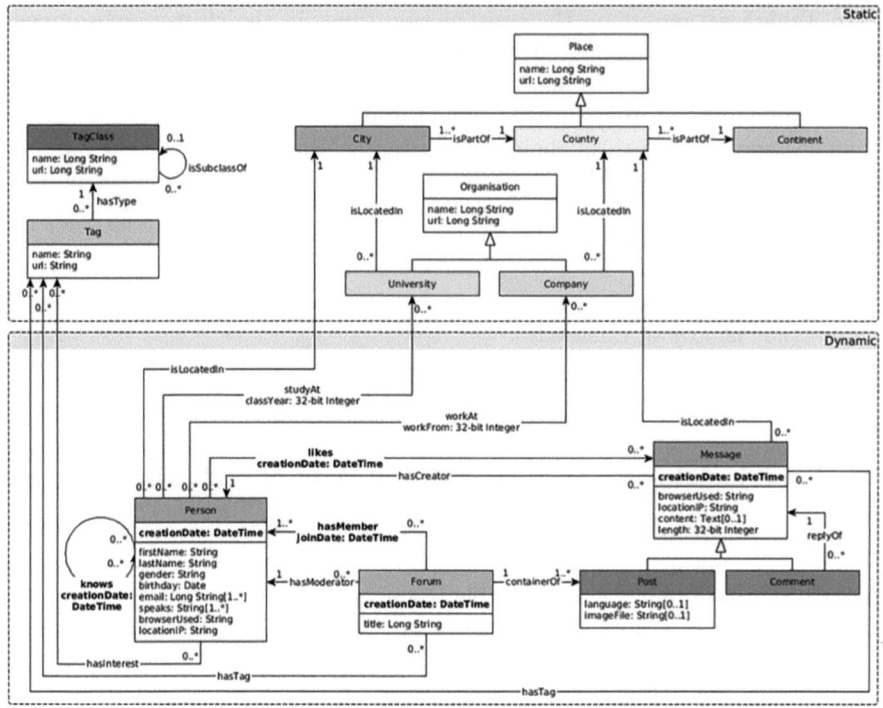

Fig. 9. The general LDBC SNB Schema.

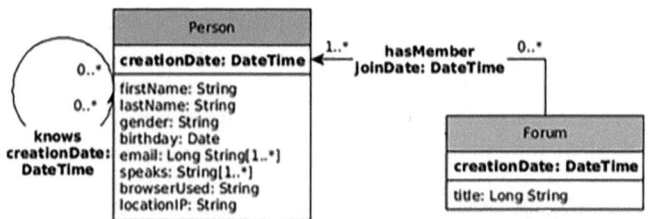

Fig. 10. The reduced LDBC SNB Schema used for our testing cases.

maximum number of values straightforwardly. Furthermore, with a more intricate endeavor, users can manipulate additional parameters. The default schema of SNB, depicted in Fig. 9, contains a static part and a dynamic part. The static graph encompasses entities that remain invariant throughout the benchmark's duration, while dynamic entities are subject to modification or removal during the benchmark execution.

The LDBC SNB offers a standardized approach for evaluating and benchmarking database systems. A key aspect of this benchmarking process involves determining scale factors (SFs) to accommodate systems of varying sizes and

Table 6. Details about the datasets used.

Name	# of vertices	# of edges
SF3	25870	668430
SF10	60800	2304951
SF3 Person and Forum	285499	10499492

resource limitations. SFs are calculated based on the ASCII size of resulting output files, measured in Gibibytes (GiB), utilizing a specific serializer as outlined in Sect. 3.4.2 of the benchmark documentation. These SFs represent different dataset sizes: SF1 equals 1 GiB, SF100 equals 100 GiB, and SF10,000 equals 10,000 GiB.

By default, all SFs span a three-year period beginning in 2010, and their calculation is based on scaling the number of individuals within the network. This means that the benchmark data generated at different SFs reflects various sizes of social networks, with SF10,000 representing a significantly larger and more intricate network compared to SF1. More information can be found in [1]

In the initial phase of our proof-of-concept benchmark, we aimed to simplify the graph structure, focusing solely on dynamic entities. Consequently, we employed a reduced schema, as illustrated in Fig. 10, comprising only of two entities: Person and Forum, along with the corresponding edges "Person Knows Person" and "Forum has Member Person". Additionally, both creationDate and deletionDate attributes are included for entities. We used the default parameters in the configuration with SF3 and SF10 sizes that are included in our reduced dataset, with only "Person" entities and "Person Knows Person" edges in "SF3" and "SF10" datasets. We also used one more extended dataset "SF3 Person and Forum" that also includes the data about "Forum" entity and "Forum has Member Person" edges. The dataset statistics are shown in Table 6.

Streaming Experiments. In a streaming environment, data typically arrives in the form of events rather than batches. Therefore, the approach used in the previous experiments that relied on node or edge loading in batches would not be suitable for this scenario. To address this limitation, we designed an event-based approach for both Cassandra and MongoDB implementations. In the context of an event-based system, the transactions represent usually simple events with minimal changes instead of making multiple data changes. For our experiments, we used the LDBC dataset generated using SF3 and SF10 with some modifications.

To establish an event-based system, we first need to identify the fundamental transactions required to represent a historical graph comprehensively. We divided the transactions into two categories: node/edge/parameter insertion and deletion. It is important to note that deletion only sets the ending time of an object since nothing is deleted in a historical graph unless explicitly stated. The insertion process requires only the start time of the object and sets the end time to infinity. The HiNode model is used to store all objects. The API was modified to accommodate the event-based approach, and some index modifications were made to enhance performance. More specifically, the following instructions are supported:

- insert_node($start$): It creates a new empty historical node v with a valid interval $[start, +\infty)$, meaning that the node is valid for all time instants after $start$. The same holds for all other operations as well, with respect to the use of the special symbol $+\infty$.
- insert_edge($u, v, start$): It creates a new edge with a valid interval $[start, +\infty)$, which must be contained in the valid interval of both nodes u and v. The same check (although not mentioned) holds for the rest of the operations.
- insert_property($p, f, val, start$): It creates a new property f in the historical object p (node or edge) with value val and a valid interval $[start, +\infty)$.
- delete_node(v, end): Node v has its valid interval shrunk since it is invalidated in the time range $(end, +\infty)$. This means that all properties of v and edges adjacent to v must be checked so that their valid intervals are still legitimate. In case they are not, then their valid intervals are shrunk.
- delete_edge(e, end): Edge e has its valid interval shrunk, since it is invalidated in the time range $(end, +\infty)$. This means that all properties of e must be checked so that their valid intervals are still legitimate. In case they are not, then their valid intervals are shrunk as well.
- delete_property(p, f, end): The valid interval of property f of the historical object p is shrunk, since it is invalidated in the time range $(end, +\infty)$.

Using combinations of these operations, we can support historical graph streaming applications.

Regarding space efficiency, the observations are the same with Sect. 5.1 with the only difference being that the SF3 Person and Forum MongoDB MT implementation, may take up to 7× more space than the corresponding Cassandra implementation. This is happening due to the preallocation of memory in MongoDB documents. In addition, one can see that there is a disagreement between Cassandra and MongoDB related to ST and MT models. While Cassandra MT uses less space that ST, for MongodDB it holds the inverse. The reason is that Cassandra is a wide-column store, while MongoDB is a document database. The former is best suited to the case where we split the columns of a table into multiple smaller ones like we did in the MT model. The latter is more efficient when the documents store more data rather than having many documents with small data each. Thus, especially in the SF datasets were we have more attributes, the latter has better results in the ST model compared to MT.

After inserting the data into each model (Cassandra/MongoDB, ST/MT), we tested both local and global queries. In particular, we employed the one-hop and the degree distribution query using the same test environment as in the previous experiments. For one-hop queries, some vertices are randomly selected and we compute the average of the queries, while for degree distribution, we query for 3 time spans, for one, two and three years (that is, for 33.3%, 66.6% and 100% of the graph respectively). The degree distribution is computed per year, meaning

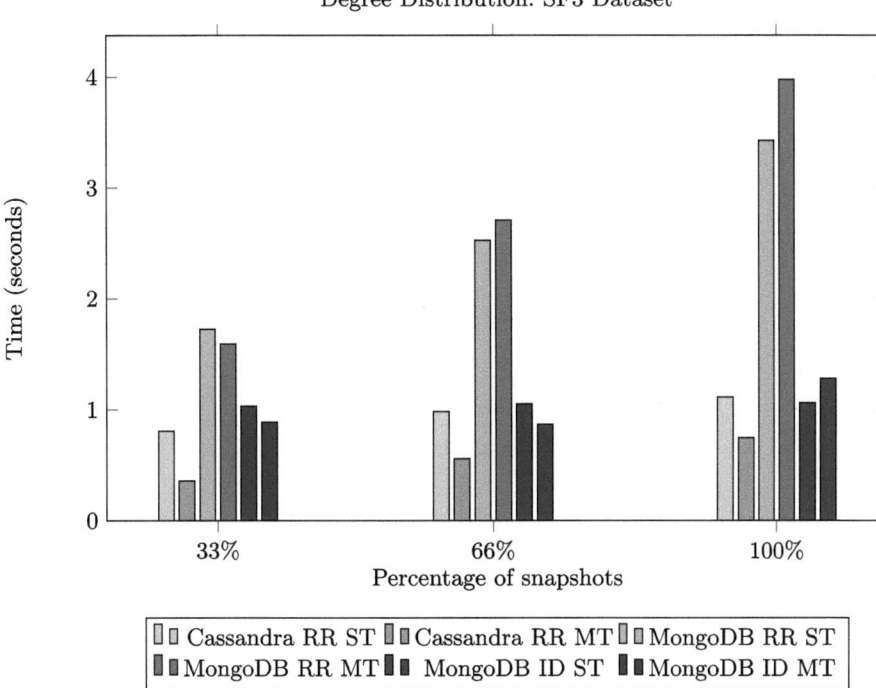

Fig. 11. Results for the vertex degree distribution query on the SF3 dataset in a single machine.

Table 7. The time to create the database from a sequence of insertions for each model and each dataset in a single machine (in minutes - $+\infty$ indicates that the construction did not complete).

		SF3	SF10	SF3 Forum
MongoDB	ST	7.20	25.75	106.22
	MT	1.72	5.93	21.458
Cassandra	ST	10.30	36.65	$+\infty$
	MT	1.745	5.762	20.331

that when we calculate the degree distribution in the time range of three years, the result will include the degree distribution for the first, second and third year separately. The models we choose to test for the MongoDB are the best ones based on the preceding experiments. The results of the experimentation in a single machine can be summarized in the following observations:

1. In Table 7, the time to build the historical graph database by consecutive insertions for various models is depicted. We can observe that the MT models have the best performance with a speedup of up to 7.2x compared to

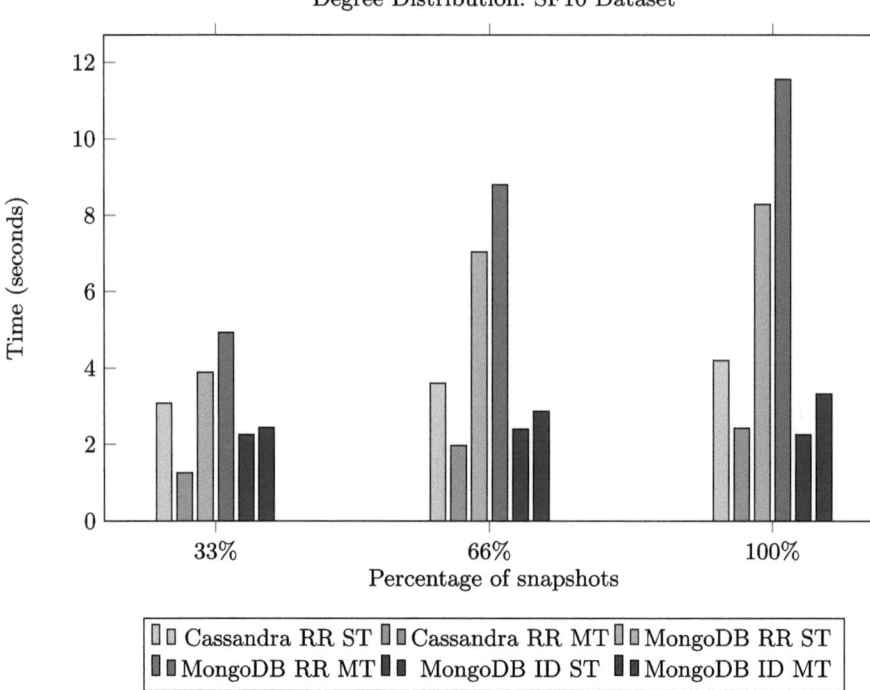

Fig. 12. Results for the vertex degree distribution query on the SF10 dataset in a single machine.

the corresponding ST models. MT models have comparable performance in Cassandra and MongoDB, but ST models have the best performance in MongoDB, outperforming the corresponding Cassandra model by up to 34.9%.

2. In Figs. 11, 12, and 13, the time needed to execute a degree distribution query is depicted for various models. For Cassandra, the best model is RR MT, which consistently outperforms its ST counterpart across all datasets. In general, Cassandra shows significant performance improvements with MT models, compared to ST models, with RR MT being up to 2.4x faster than RR ST in some cases Conversely, in MongoDB, ST models generally outperform MT models, especially when using more percentage of snapshots and when larger datasets are used. Especially, the ID ST model performs best, showing stable performance across different snapshot percentages.

3. Cassandra MT emerges as the superior model, showcasing significant performance advantages over MongoDB in most cases. For the SF3 dataset, Cassandra MT RR outperforms MongoDB ST ID by 29.80% –65.51%, translating to a speedup of up to 3.1×. These performance gains are consistent across both SF3 and SF10 datasets, in the later dataset Cassandra MT RR improves the performance by up to 44.35% in 33% and 66% while when we query for 100% of the snapshots MongoDB ST ID is the best-performing model improving the

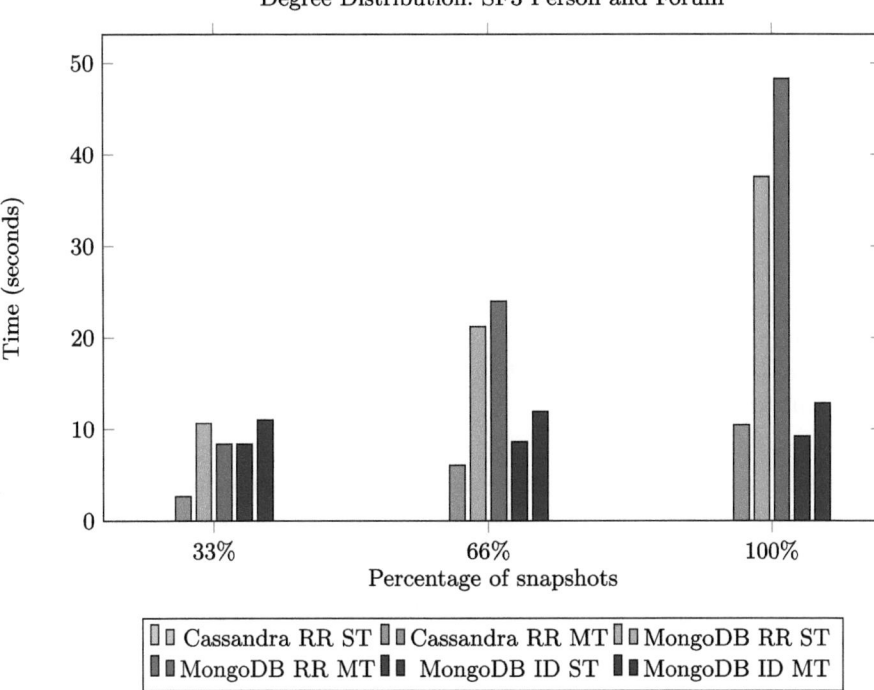

Fig. 13. Results for the vertex degree distribution query on the SF3 Person and Forum dataset in a single machine.

performance by 7.42% compared to the best Cassandra model. Notably, the ID processing query mode in MongoDB demonstrates exceptional stability, with execution time increases limited to at most 15% as the query percentage of snapshots grows. This contrasts sharply with other modes and models, particularly Cassandra, where time increases can reach up to 100% in the SF3 Person and Forum dataset as snapshot percentages increase.

4. The MongoDB ST ID model is becoming more efficient compared to other models as the size of the dataset increases. The size of the datasets can be seen in Table 6, although the size needed in its database to store them is depicted in Table 8. Moving from SF3 to SF10 we increased more than 3.5 times the number of edges while we can observe for Cassandra ST and MT models a 3.7 and 3.4 times increase of time spent in the query, while for MongoDB the time increase is 2.2 and 3 times for ST ID and MT ID models respectively, showing better scale while data increases. Between SF10 and SF3 Person and Forum, the edges are increased 4.5 times, while the query time increased 3.7 times for the MongoDB ST ID model, 4.14 for MongoDB MT ID model, and 2–4 times for Cassandra MT model.

Table 8. Size of the database for each model and each dataset respectively.

		SF3	SF10	SF3 Extended
MongoDB	ST	132.7 MB	437.6 MB	1.9 GB
	MT	145.3 MB	479.6 MB	2.22 GB
CASSANDRA	ST	62.5 MB	278 MB	N/A
	MT	20 MB	108 MB	336.7 MB

5. Cassandra has an advantage because it exhibits high-throughput transactions, but because of the nature of the schema, it will get setbacks when multiple updates happen in the parameters or edges (which are not considered in the current test cases).

Table 9. Construction time (a sequence of insertions) comparison for SF3, SF10, and the extended SF3 datasets in the cluster (in minutes - N/A indicates that the construction was not carried out for this dataset for implementation reasons).

		SF3	SF10	Extended SF3
Cassandra	ST	103.3	373.4	N/A
	MT	22.2	72.6	270.1
MongoDB	ST	251.3	906.4	2703.9
	MT	143.2	268.8	1055.1

We executed both MongoDB and Cassandra models in the cluster and the results for Degree Distribution Query can be seen in Figs. 14, 15 and 16, while for one hop query can be seen in Figs. 17, 18 and 19. Also in Table 9, is depicted the times to insert the data into the cluster. The observations, with some exceptions, are similar to the snapshot-based experiments:

1. For Cassandra, the best-performing models are MT. In terms of degree distribution, they manage to reduce execution time by 65.4% to 75.7%, while for one-hop queries, the reduction ranges from 11% to 50%.
2. For MongoDB, the best-performing models differ based on the application. For the global query degree distribution, ID ST is the best-performing model, managing to reduce execution time by 29.6% up to 35.7%. However, for the case of one-hop local query, the best performing model is MT, reducing the execution time by 2.5% up to 20.7%.
3. Comparing the best Cassandra and MongoDB models, the dominant model differs on the application, size of the dataset or query percentage.
4. On the one-hop query, except when running for 100% of the graph in SF3 and SF10, the best-performing model is Cassandra MT, managing to reduce execution time by 5.5% up to 45.8% compared to the best-performing MongoDB model. At 100% of the graph, MongoDB ST is the dominant model,

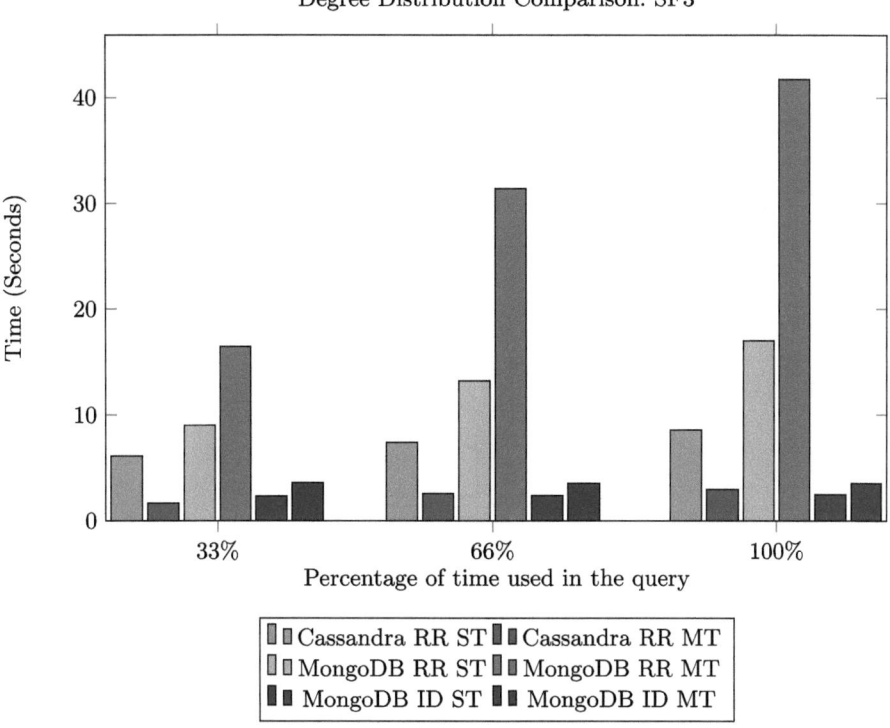

Fig. 14. Degree distribution query comparison for the SF3 dataset in the cluster.

managing to reduce the execution time by 58% in SF3 and 19.9% in SF10 when compared to the best Cassandra model.

5. Regarding the degree distribution query, in SF3 and SF10, Cassandra MT is the best-performing model while querying for 33% of the graph, reducing the execution time by 39.6% and 45% compared to the best MongoDB model. For degree distribution queries that involve 66% or 100% of the graph, MongoDB ST ID is the best-performing model, managing to reduce the execution time by up to 16%. In the extended SF3 dataset, the best-performing model is Cassandra MT (when it can be executed) managing to reduce the execution time between 27.8% up to 63.8% compared to the best MongoDB alternative. It can be observed that while doing so, the percentage is being reduced while we query a larger percentage of time.

5.3 Discussion

Regarding the differences between Cassandra and MongoDB implementations we observe the following:

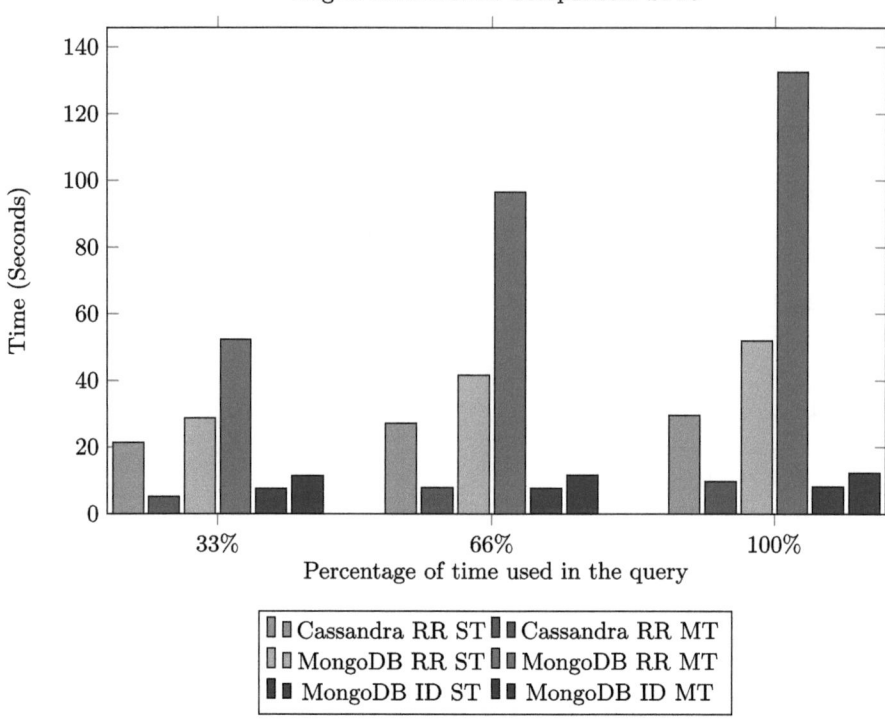

Fig. 15. Degree distribution query comparison for the SF10 in the cluster.

1. *Local Queries in Snapshots*: Cassandra is more efficient (but not by far) than MongoDB except when a large number of time instants is involved in the query. In this case, MongoDB is more efficient.
2. *Global Queries in Snapshots*: In cluster mode, MongoDB is superior to Cassandra when the ID approach is adopted. In the other cases and in local mode the results are mixed.
3. *Transactions*: MongoDB using ST is faster than Cassandra both in local and in cluster mode.
4. *Streaming*: Both in local and cluster mode, Cassandra and MongoDB have similar performance. For global queries, MongoDB with ID seems to be the best choice, while for local queries the results are mixed.
5. In general, ST is best performing in MongoDB, while MT in Cassandra.

Regarding the comparison of Single Table (ST) model that corresponds to a pure-vertex centric model, and the Multiple Table (MT) model in which a diachronic node has been split into a small number of collections, we reach the following conclusions related to MongoDB:

1. *Local Queries in Snapshots:* In a cluster environment, our small set of local queries shows that ST is slightly better than MT.

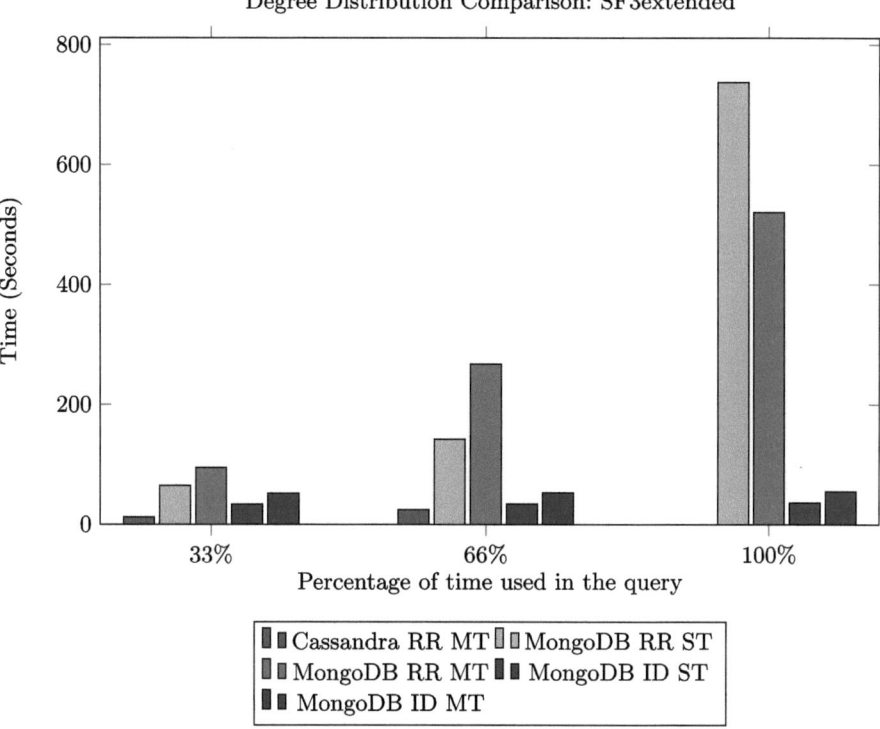

Fig. 16. Degree distribution query comparison for the extended SF3 dataset in the cluster.

2. *Global Queries in Snapshots:* Both in a local machine and in a cluster environment, ST and MT have similar performance, perhaps with a slight improvement from the ST side in the cluster environment for the ID approach.
3. *Transactions and Streaming*: The ST model seems to cope better in a cluster environment (this is a difference when compared to the local machine environment) when compared to MT for transactions.
4. *Streaming Queries*: Looking at the experiments on the cluster, the ST model seems to be doing a little better in global queries while in local queries the MT model seems to be doing better.
5. ST has slightly better space usage than MT. However, this depends largely on how fine is the splitting to multiple tables for the MT model, and the difference can become quite larger for finer (more tables) splitting.

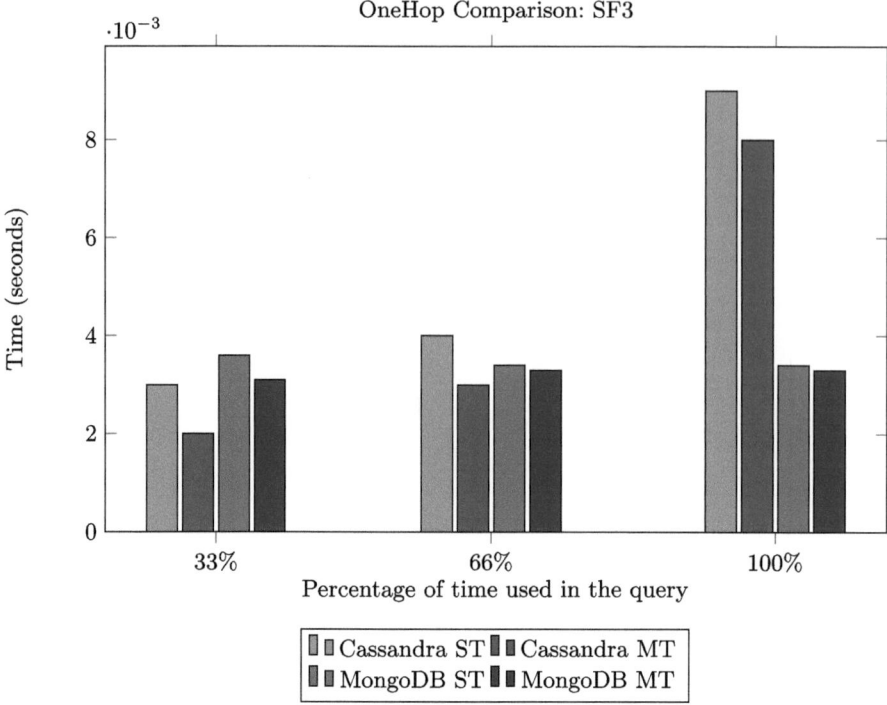

Fig. 17. OneHop query comparison for the SF3 dataset in the cluster.

All in all, there is no clear winner - and we did not expect for one - since the performance depends on the type of query, on the size of the query time interval, as well as on the dataset. However, the goal of this paper was to show that the vertex-centric approach in a NoSQL environment has certain merits even in the case of global queries. One can extend this work among several axes. One can look at different queries and come up with mixed workloads (OLTP and OLAP) of various types. In addition, a side-effect that we discovered during our research, is the need for a generator for historical graph workloads. Our great aspiration is to make an existing graph database incorporate the notion of time as a first-class citizen. The current research was targeted towards this goal regarding the storage model and a certain type of queries.

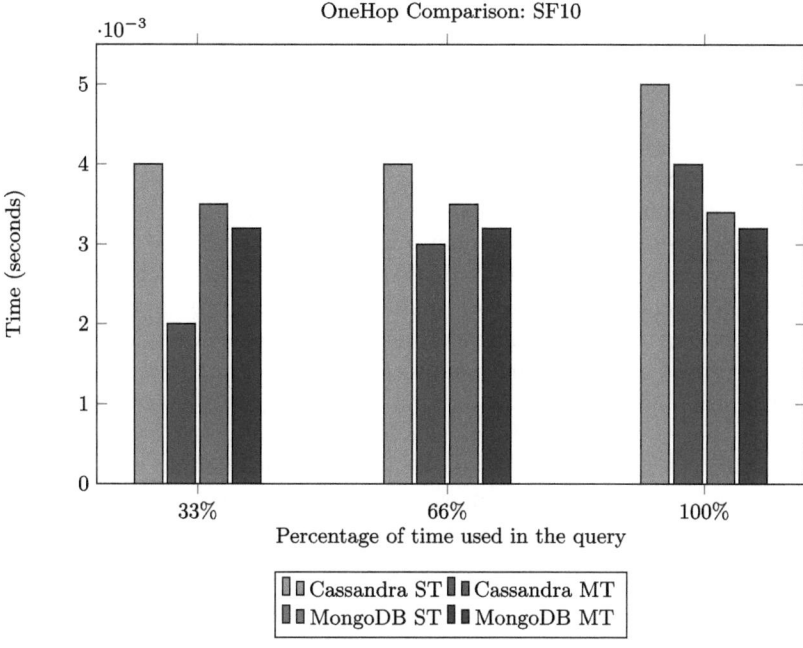

Fig. 18. OneHop query comparison for the SF10 dataset in the cluster.

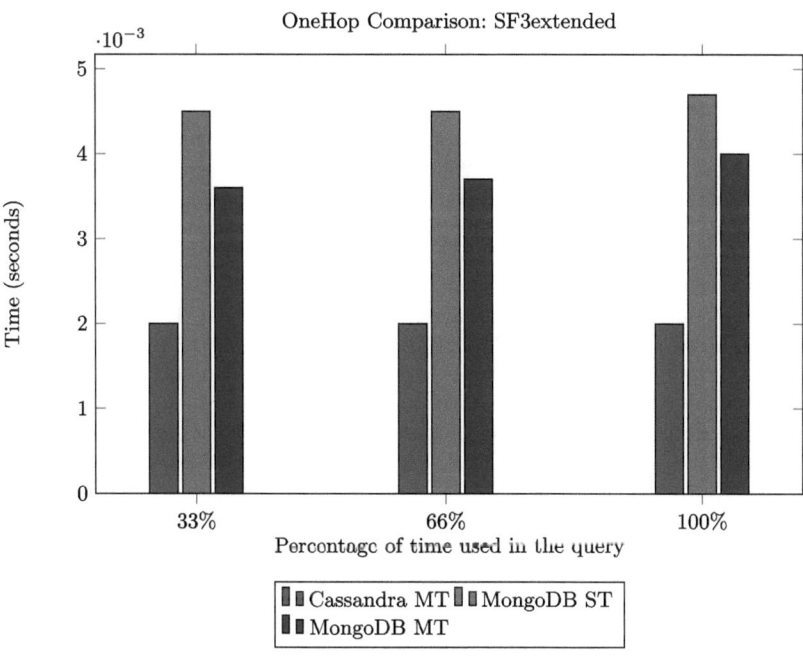

Fig. 19. OneHop query comparison for the extended SF3 dataset in the cluster.

6 Conclusion

In this work, we have shown how the HiNode vertex-centric approach to storing time-varying graphs can be implemented in MongoDB, so that significant improvements in global queries can be achieved compared to the previous NoSQL-based implementation (over 4× in some cases), while we also implemented the streaming functionality in both our systems. We focused mainly on global queries, since this is the main expected bottleneck of a vertex-centric approach. We made extensive comparisons between Cassanda and MongoDB implementations of different flavors, highlighting in the process some differences when considering a local machine environment or a cluster environment. In the process, the need arose to create appropriate datasets and set the stage for a future extension of known generators for historical graphs and their workloads.

We test successfully the Hinode model (a vertex-centric model) in various cases including streaming scenarios, transactions, demanding a partial vertex-centric approach (MT) and comparing it with a full vertex centric approach (ST), snapshots and time-interval based representations. In addition, we tested the Hinode model in the case where the schema of the graph can change with the addition of new properties in the nodes/edges (by using the extended SF3 dataset).

References

1. The LDBC social network benchmark (version 2.2.4-SNAPSHOT, commit 9005c16). https://ldbcouncil.org/ldbc_snb_docs/ldbc-snb-specification.pdf
2. Andriamampianina, L., Ravat, F., Song, J., Vallès-Parlangeau, N.: A generic modelling to capture the temporal evolution in graphs. In: 16e journées EDA : Business Intelligence & Big Data (EDA 2020), vol. RNTI-B-16, pp. 19–32. Lyon, France (2020). https://hal.archives-ouvertes.fr/hal-03109670
3. Angles, R., et al.: The LDBC social network benchmark (2024)
4. Bichl, J., Driessen, T., Langermeier, M., Bauer, B.: GraphVault: a temporal graph persistence engine. In: Proceedings of the 26th International Conference on Enterprise Information Systems - Volume 1: ICEIS, pp. 224–231. INSTICC, SciTePress (2024). https://doi.org/10.5220/0012556500003690
5. Bok, K., Kim, G., Lim, J., Yoo, J.: Historical graph management in dynamic environments. Electronics 9(6) (2020). https://doi.org/10.3390/electronics9060895
6. Byun, J.: Enabling time-centric computation for efficient temporal graph traversals from multiple sources. IEEE Trans. Knowl. Data Eng. (2020). https://doi.org/10.1109/TKDE.2020.3005672
7. Byun, J., Woo, S., Kim, D.: Chronograph: enabling temporal graph traversals for efficient information diffusion analysis over time. IEEE Trans. Knowl. Data Eng. 32(3), 424–437 (2020). https://doi.org/10.1109/TKDE.2019.2891565
8. Casteigts, A., Flocchini, P., Quattrociocchi, W., Santoro, N.: Time-varying graphs and dynamic networks. Int. J. Parallel Emergent Distrib. Syst. 27(5), 387–408 (2012)
9. Christ, L., Gomez, K., Rahm, E., Peukert, E.: Distributed graph pattern matching on evolving graphs (2020)

10. Dhulipala, L., Blelloch, G.E., Shun, J.: Low-latency graph streaming using compressed purely-functional trees. In: Proceedings of the 40th ACM SIGPLAN Conference on Programming Language Design and Implementation, pp. 918–934. PLDI 2019, Association for Computing Machinery, New York, NY, USA (2019)
11. Ding, M., Yang, M., Chen, S.: Storing and querying large-scale spatio-temporal graphs with high-throughput edge insertions. arXiv preprint arXiv:1904.09610 (2019)
12. Driscoll, J.R., Sarnak, N., Sleator, D.D., Tarjan, R.E.: Making data structures persistent. J. Comput. Syst. Sci. **38**(1), 86–124 (1989). https://doi.org/10.1016/0022-0000(89)90034-2
13. Eguiluz, V.M., Zimmermann, M.G., Cela-Conde, C.J., Miguel, M.S.: Cooperation and the emergence of role differentiation in the dynamics of social networks. Am. J. Sociol. **110**(4), 977–1008 (2005)
14. Gandhi, S., Simmhan, Y.: An interval-centric model for distributed computing over temporal graphs. In: 2020 IEEE 36th International Conference on Data Engineering (ICDE), pp. 1129–1140 (2020). https://doi.org/10.1109/ICDE48307.2020.00102
15. Gedik, B., Bordawekar, R.: Disk-based management of interaction graphs. IEEE Trans. Knowl. Data Eng. **26**(11), 2689–2702 (2014). https://doi.org/10.1109/TKDE.2013.2297930
16. Gonzalez, J.E., Low, Y., Gu, H., Bickson, D., Guestrin, C.: PowerGraph: distributed graph-parallel computation on natural graphs, pp. 17–30. OSDI2012, USENIX Association (2012)
17. Han, W., Li, K., Chen, S., Chen, W.: Auxo: a temporal graph management system. Big Data Min. Anal. **2**(1), 58–71 (2019). https://doi.org/10.26599/BDMA.2018.9020030
18. Han, W., et al.: Chronos: a graph engine for temporal graph analysis. In: Proceedings of the Ninth European Conference on Computer Systems. EuroSys 2014, Association for Computing Machinery, New York, NY, USA (2014). https://doi.org/10.1145/2592798.2592799,
19. Hartmann, T., Fouquet, F., Jimenez, M., Rouvoy, R., Le Traon, Y.: Analyzing complex data in motion at scale with temporal graphs (2017). https://doi.org/10.18293/SEKE2017-048
20. Hou, J., et al.: AEONG: an efficient built-in temporal support in graph databases. Proc. VLDB Endow. **17**(6), 1515–1527 (2024). https://doi.org/10.14778/3648160.3648187,
21. Huang, H., Song, J., Lin, X., Ma, S., Huai, J.: TGraph: a temporal graph data management system. In: Proceedings of the 25th ACM International on Conference on Information and Knowledge Management, pp. 2469–2472. CIKM 2016, Association for Computing Machinery, New York, NY, USA (2016). https://doi.org/10.1145/2983323.2983335,
22. Iyer, A.P., Li, L.E., Das, T., Stoica, I.: Time-evolving graph processing at scale. In: Proceedings of the Fourth International Workshop on Graph Data Management Experiences and Systems, pp. 1–6 (2016)
23. Iyer, A.P., Pu, Q., Patel, K., Gonzalez, J.E., Stoica, I.: TEGRA: efficient ad-hoc analytics on evolving graphs. In: 18th USENIX Symposium on Networked Systems Design and Implementation (NSDI 21), pp. 337–355. USENIX Association (2021). https://www.usenix.org/conference/nsdi21/presentation/iyer
24. Ju, X., Williams, D., Jamjoom, H., Shin, K.G.: Version traveler: fast and memory-efficient version switching in graph processing systems. In: 2016 USENIX Annual Technical Conference (USENIX-ATC 16), pp. 523–536 (2016)

25. Junghanns, M., Petermann, A., Teichmann, N., Gómez, K., Rahm, E.: Analyzing extended property graphs with apache flink. In: Proceedings of the 1st ACM SIGMOD Workshop on Network Data Analytics. NDA 2016, Association for Computing Machinery, New York, NY, USA (2016). https://doi.org/10.1145/2980523.2980527,
26. Khurana, U., Deshpande, A.: Efficient snapshot retrieval over historical graph data. In: 29th IEEE International Conference on Data Engineering, ICDE 2013, Brisbane, Australia, 8-12 April 2013, pp. 997–1008 (2013)
27. Khurana, U., Deshpande, A.: Efficient snapshot retrieval over historical graph data. In: 2013 IEEE 29th International Conference on Data Engineering (ICDE), pp. 997–1008 (2013). https://doi.org/10.1109/ICDE.2013.6544892
28. Khurana, U., Deshpande, A.: Storing and analyzing historical graph data at scale. In: EDBT, pp. 77–88 (2016)
29. Khurana, U., Deshpande, A.: Storing and analyzing historical graph data at scale. In: Pitoura, E., Maabout, S., Koutrika, G., Marian, A., Tanca, L., Manolescu, I., Stefanidis, K. (eds.) Proceedings of the 19th International Conference on Extending Database Technology, EDBT 2016, Bordeaux, France, 15-16 March 2016, pp. 65–76 (2016). OpenProceedings.org (2016). https://doi.org/10.5441/002/edbt.2016.09,
30. Kosmatopoulos, A., Giannakopoulou, K., Papadopoulos, A.N., Tsichlas, K.: An overview of methods for handling evolving graph sequences. In: Karydis, I., Sioutas, S., Triantafillou, P., Tsoumakos, D. (eds.) ALGOCLOUD 2015. LNCS, vol. 9511, pp. 181–192. Springer, Cham (2016). https://doi.org/10.1007/978-3-319-29919-8_14
31. Kosmatopoulos, A., Gounaris, A., Tsichlas, K.: HiNode: implementing a vertex-centric modelling approach to maintaining historical graph data. Computing **101**(12), 1885–1908 (2019). https://doi.org/10.1007/s00607-019-00715-6
32. Kosmatopoulos, A., Tsichlas, K., Gounaris, A., Sioutas, S., Pitoura, E.: HiNode: an asymptotically space-optimal storage model for historical queries on graphs. Distrib. Parallel Databases **35**(3–4), 249–285 (2017). https://doi.org/10.1007/s10619-017-7207-z
33. Kumar, P., Huang, H.H.: Graph one: a data store for real-time analytics on evolving graphs. ACM Trans. Storage **15**(4) (2020). https://doi.org/10.1145/3364180,
34. Labouseur, A.G., et al.: The g* graph database: efficiently managing large distributed dynamic graphs. Distrib. Parallel Databases **33**(4), 479–514 (2015)
35. Leskovec, J., Krevl, A.: Snap datasets: high-energy physics phenomenology citation network (2014). https://snap.stanford.edu/data/cit-HepPh.html. dataset originally released as part of the 2003 KDD Cup. Covers citations among arXiv HEP-PH papers from Jan 1993 to Apr 2003
36. Leskovec, J., Krevl, A.: Snap datasets: high-energy physics theory citation network (2014). https://snap.stanford.edu/data/cit-HepTh.html. dataset originally released as part of the 2003 KDD Cup. Covers citations among arXiv HEP-TH papers from Jan 1993 to Apr 2003
37. Leskovec, J., Krevl, A.: Snap datasets: US patent citation network (2014). https://snap.stanford.edu/data/cit-Patents.html. dataset originally released by the National Bureau of Economic Research (NBER). Covers citations among US utility patents granted from 1963 to 1999, with citation data from 1975 to 1999
38. Lightenberg, W., Pei, Y., Fletcher, G., Pechenizkiy, M.: TINK: a temporal graph analytics library for apache Flink. In: Companion Proceedings of the The Web Conference 2018, pp. 71–72 (2018)

39. Lim, S., Coy, T., Lu, Z., Ren, B., Zhang, X.: NVGraph: enforcing crash consistency of evolving network analytics in NVMM systems. IEEE Trans. Parallel Distrib. Syst. **31**(6), 1255–1269 (2020). https://doi.org/10.1109/TPDS.2020.2965452
40. Maduako, I., Wachowicz, M., Hanson, T.: STVG: an evolutionary graph framework for analyzing fast-evolving networks. J. Big Data **6**(1), 1–24 (2019)
41. Massri, M., Raipin Parvedy, P., Meye, P.: GDBAlive: a temporal graph database built on top of a columnar data store. J. Adv. Inf. Technol. **12** (2020). https://doi.org/10.12720/jait.12.3.169-178
42. Miao, Y., et al.: ImmortalGraph: a system for storage and analysis of temporal graphs. ACM Trans. Storage **11**(3) (2015). https://doi.org/10.1145/2700302
43. Michail, O., Spirakis, P.G.: Elements of the theory of dynamic networks. Commun. ACM **61**(2), 72 (2018)
44. Moffitt, V.Z.: Framework for querying and analysis of evolving graphs. Ph.D. thesis (2017). https://doi.org/10.13140/RG.2.2.16079.64166, https://www.proquest.com/docview/1946186055?pq-origsite=gscholar&fromopenview=true
45. Moffitt, V.Z., Stoyanovich, J.: Portal: a query language for evolving graphs. CoRR abs/1602.00773 (2016). http://arxiv.org/abs/1602.00773
46. Moffitt, V.Z., Stoyanovich, J.: Towards sequenced semantics for evolving graphs. In: EDBT, pp. 446–449 (2017)
47. Mondal, J., Deshpande, A.: Managing large dynamic graphs efficiently. In: Proceedings of the 2012 ACM SIGMOD International Conference on Management of Data, pp. 145–156. SIGMOD 2012 (2012)
48. Raimundo, R.L., Guimarães, P.R., Evans, D.M.: Adaptive networks for restoration ecology. Trends Ecol. Evol. **33**, 664–675 (2018). https://doi.org/10.1016/j.tree.2018.06.002
49. Ramesh, S., Baranawal, A., Simmhan, Y.: Granite: a distributed engine for scalable path queries over temporal property graphs. J. Parallel Distrib. Comput. **151**, 94–111 (2021). https://doi.org/10.1016/j.jpdc.2021.02.004
50. Ren, C., Lo, E., Kao, B., Zhu, X., Cheng, R.: On querying historical evolving graph sequences. PVLDB **4**(11), 726–737 (2011)
51. Rost, C., et al.: Distributed temporal graph analytics with GRADOOP. VLDB J. **31**(2), 375–401 (2021). https://doi.org/10.1007/s00778-021-00667-4
52. Rost, C., Thor, A., Rahm, E.: Analyzing temporal graphs with gradoop. Datenbank-Spektrum **19**(3), 199–208 (2019)
53. Sahu, S., Salihoglu, S.: Graphsurge: graph analytics on view collections using differential computation. In: Proceedings of the 2021 International Conference on Management of Data, pp. 1518–1530 (2021)
54. Saidani, S.: Self-Reconfigurable Robots Topodynamic, vol. 3, pp. 2883–2887 (2004)
55. Shao, B., Wang, H., Li, Y.: Trinity: a distributed graph engine on a memory cloud. In: Proceedings of the ACM SIGMOD International Conference on Management of Data, SIGMOD 2013, pp. 505–516 (2013)
56. Smith, D.M.D., Onnela, J.P., Lee, C.F., Fricker, M.D., Johnson, N.F.: Network automata: coupling structure and function in dynamic networks. Adv. Complex Syst. **14**(03), 317–339 (2011)
57. Spillane, S.R., et al.: A demonstration of the G_* graph database system. In: 29th IEEE International Conference on Data Engineering, ICDE 2013, Brisbane, Australia, April 8-12, 2013, pp. 1356–1359 (2013)
58. Spitalas, A., Gounaris, A., Tsichlas, K., Kosmatopoulos, A.: Investigation of database models for evolving graphs. In: Combi, C., Eder, J., Reynolds, M. (eds.) 28th International Symposium on Temporal Representation and Reasoning, TIME

2021, September 27-29, 2021, Klagenfurt, Austria. LIPIcs, vol. 206, pp. 6:1–6:13. Schloss Dagstuhl - Leibniz-Zentrum für Informatik (2021). https://doi.org/10.4230/LIPIcs.TIME.2021.6
59. Spitalas, A., Tsichlas, K.: MAGMA: proposing a massive historical graph management system. In: Foschini, L., Kontogiannis, S. (eds.) Algorithmic Aspects of Cloud Computing, pp. 42–57. Springer International Publishing, Cham (2023). https://doi.org/10.1007/978-3-031-33437-5_3
60. Steer, B., Cuadrado, F., Clegg, R.: Raphtory: streaming analysis of distributed temporal graphs. Futur. Gener. Comput. Syst. **102**, 453–464 (2020). https://doi.org/10.1016/j.future.2019.08.022
61. Theodorakis, G., Clarkson, J., Webber, J.: AION: efficient temporal graph data management. In: EDBT, pp. 501–514 (2024)
62. Vijitbenjaronk, W.D., Lee, J., Suzumura, T., Tanase, G.: Scalable time-versioning support for property graph databases. In: 2017 IEEE International Conference on Big Data (Big Data), pp. 1580–1589 (2017). https://doi.org/10.1109/BigData.2017.8258092

Data Assetization Journey: Concepts, Principles, and Illustrations

Zakaria Maamar[1](✉), Belkacem Chikhaoui[2], Amel Benna[3],
Vanilson Burégio[1], and Djamal Benslimane[4]

[1] University of Doha for Science and Technology, Doha, Qatar
{zakaria.maamar,vanilson.buregio}@udst.edu.qa
[2] Applied Artificial Intelligence Institute, TELUQ University, Montreal, Canada
belkacem.chikhaoui@teluq.ca
[3] Research Center for Scientific and Technical Information, Algiers, Algeria
abenna@cerist.dz
[4] Lyon 1 University, Lyon, France
djamal.benslimane@univ-lyon1.fr

Abstract. In line with "data is the new oil", this paper presents a journey for a successful transformation of data into assets. The journey consists of 3 stopovers featuring each particular operations to perform. The eligibility stopover ensures that data is "worth" assetizing since not all data can become assets. Next, the enrichment stopover provides extra details about data allowing a better "reasoning" over these data after passing the eligibility check successfully. Finally, the governance stopover develops policies for managing data as an asset in terms of depreciation, transferability, disposability, and convertibility. Policies are specified in the Open Digital Rights Language (ODRL) to express permission, prohibition, and obligation rules that steer the data assetization journey. A system demonstrating the technical doability of this assetization with emphasis on the eligibility stopover is also presented in the paper.

Keywords: Asset · Data · Eligibility · Enrichment · Governance · Policy

1 Introduction

Over the years, data has become a commodity fueling the economies of many countries to the extent that global data creation is projected to reach 180 zettabyte by 2025[1]. This massive volume of data will definitely put pressure on the world's ICT infrastructure when time comes to collect data, safeguard data, communicate data, process data, to cite just some. Aligning ourselves with the prevalent practice of Anything-as-a-Service (∗aaS), we recently advocated for Data-as-an-Asset (DaaA) where data is elevated to become an asset [20]. To

---

[1] https://www.statista.com/statistics/871513/worldwide-data-created/#statisticContainer.

encourage DaaA adoption, this paper presents a data assetization journey. To this end, we shed light on first, how both data's intrinsic features (e.g., freshness, confidentiality, and shareability) and asset's intrinsic features (e.g., depreciation, disposability, and convertibility) shape the journey and second, what steps, (AI) techniques, and modules are needed to complete the journey. Although the prevalent practice in the ICT community is Data-as-a-Service (DaaS) where data is provisioned for use to third parties [25], this practice has a limited emphasis on how data evolves over time from generation and collection to processing, distribution, and sometimes disposal. Data is a "living" element that many countries are considering as another source of income; *"data is the new oil"*[2]. To address this limited emphasis, we propose policies to "steer" data evolution and use the Open Digital Rights Language (ODRL) to specify policies. ODRL provides relevant constructs (e.g., policy, rule, and action) for managing (mainly digital) assets based on permission, prohibition, and duty rules [7,30]. Some immediate benefits of having ODRL policies "steer" data evolution and thus, the assetization journey include the definition of what is permitted to exercise over data, what is prohibited from exercising over data, and what must be exercised over data. In addition, coupling ODRL with other techniques such as AI would allow the data assetization journey to tap into machine learning for pattern detection and federated learning for privacy preservation, for example.

For a successful data assetization journey, we define in this paper the main features of data to illustrate for instance, how freshness could make an asset linked to a data depreciate very quickly and how integrity could also make an asset linked to another data be disposed of very quickly, too. Indeed, depreciation, transferability, disposability, and convertibility features are critical for asset management. Over time, an asset can depreciate becoming unnecessary, can be transferred ending up with a new owner and/or in a new location, can be converted becoming a source of income, and can be disposed of making room for another asset. To what extent data's and asset's features impact each other should be examined to understand how these features either expedite or slow down the assetization journey.

In this paper, we present a *data assetization journey* consisting of 3 stopovers, *eligibility, enrichment,* and *governance*. In each stopover, specific operations are performed, so that data becomes an asset. Our contributions are (i) definition of assetization from a data perspective, (ii) impact analysis of data's and asset's features on assetization, (iii) identification of steps that achieve data assetization, (iv) specification of ODRL policies to "steer" the data assetization journey, and (v) development of a proof-of-concept implementing the data assetization journey. The rest of this paper is organized as follows. Section 2 presents some related works and a case study. Section 3 details the data assetization journey in terms of data and asset features and the journey's stopovers. An implementation of one of the stopovers is discussed in Sect. 4. Prior to concluding and listing some future work elements in Sect. 6, Sect. 5 analyzes the data assetization journey.

---

[2] https://www.sheffield.ac.uk/cs/people/academic-visitors/clive-humby.

2 Background

This section presents some works related to assetization from a broad perspective and then, suggests a case study used for illustration purposes.

2.1 Related Work

According to Birch and Muniesa, an asset is *"something that can be owned or controlled, traded, and capitalized as a revenue stream, often involving the valuation of discounted future earnings in the present"* [5]. Referring to the same authors, asset creation also enables organizations to have a sustainable economic income in contrast to the final sale of a product or the final provisioning of a service [4]. Unfortunately, Birch and Muniesa do not provide any guideline about the best way to create assets. In term of accessibility, data as an asset can either be open being available to all potential users, be closed being available only to a certain organization, typically the one that collected or produced it, or be shared being available to multiple organizations, though it is not openly available [10]. On top of accessibility, data is useful for forecasting during planning, operating real-time devices, and reflecting on historic events, and sharing [10].

In [2], Beauvisage and Mellet discuss the monetary values of some personal data based on certain past studies. For instance, age and weight were valued at \$57 and \$74, respectively, according to [13], and browsing history was valued at €7 according to [8]. To collect these values using a Web plugin like in [8], users were immersed in fictitious situations in which they had to weigh up the cost of disclosing information about themselves (e.g., age, income, and browsing history) against financial incentives (e.g., discount and revenue). Beauvisage and Mellet came to the conclusion that data assetization is a process combining capture and re-purposing, can be considered as an entrepreneurial work, and can lead to plural and versatile products. Such a process could enhance our data assetization journey mainly for establishing data values.

In [4], Birch et al. list domains and applications where assetization is used including science and technology (knowledge transformation and production into intellectual property), human geography (agricultural technologies and carbon finance analysis), and critical data studies (analysis of personal, urban, and health data). In the agriculture domain, Hackfort et al. mention that many organizations appear to shifting their business practices to focus on data as a key asset and driver of profit [12]. Data is being "harvested" and aggregated into large-scale datasets. For a better appreciation of data as an asset in the agriculture domain, Hackfort et al. raised the question of *"how is agricultural data transformed into value by the most powerful agribusinesses and ag-tech firm"* [12]. To address this question, 3 strategies were developed: use data to have farmers locked into relationships with particular firms, use big data analytics for price setting and hedging, and use big data internally to drive product development and for targeted marketing. While these strategies could be packaged into a process to follow, how to identify data to assetize, how to assetize data, and how to derive profit from data are unaddressed questions.

In [9], Chen defines data assets as data generated or acquired by organizations through production, operational, or transactional processes. These data assets exhibit multiple characteristics such as high universality, non-depletion, dynamism, and diverse forms of value creation. Organizations are expected to possess ownership or usage rights over data assets and adhere to legal regulations and agreements. Chen categorizes data assets into own-use and transactional allowing to improve the information efficiency of the capital market.

In [15], Kang et al. shed light on problems hindering data assetization and hence, data circulation and value release in China. Particular problems include lack of standards for both identification of data assets and unification of data types and products across industries, lack of laws on data ownership, lack of data asset quality assessment techniques, lack of a standardized and mature data asset value evaluation system, and lack of national or industry standards for data property registration. To address these problems, Kang et al. suggest a data assetization path consisting of data resource utilization (data acquisition, transmission, aggregation, governance, storage, and rough processing), resource productization (research and development of data products), and data assetization (data-asset identification, data-asset ownership confirmation, data-asset quality assessment, data-asset value assessment, and data-asset registration) Although Kang et al.'s path and our journey have the same purpose, we are different in the sense of ensuring that data is worth assetization prior to taking any action. We will later show that not all data can be assetized.

In [31], Wang et al. emphasize that data is not inherently asset-oriented. To become an asset (i.e., eligible as per our terminology, Fig. 2), data must meet the characteristics of an asset itself and generate value for the organization. First, data assets must be quantifiable, so they are identifiable. Second, data assets can create value, so they bring economic benefits to the organization. Third, data assets have a life cycle, so they are tracked since their values gradually decrease over time leading to their deactivation and disposal. As part of their research in China's grid data asset management, Wang et al. also presented a method of data asset management based on data management body of knowledge, data-management maturity model, and "9+2" model of data asset management.

In [34], Xu et al. conducted a systematic literature review of the concept of data assets and how organizations can assetize their data. They raised 3 research questions namely, what are the evolution and controversies of the concept of data assets, how to define data assets in the context of the digital economy, and what are the strategic imperatives for enterprises to achieve data assetization? Along with these questions, Xu et al. developed a data assetization framework that encompasses 3 steps referred to as data resource investment, data capability building, and data application. On the one hand, resource encompasses existing human, technical, and data-related assets that organizations use to effectively analyze and leverage data assets. On the other hand, capability is about problem-solving proficiencies that organizations develop to overcome the obstacles and tackle the challenges encountered during data assetization. Similarly to the respective works of Chen [9], Kang et al., [15], and Wang et al. [31],

Xu et al.'s framework does not cover critical aspects we deem necessary in data assetization namely, eligibility, enrichment, and governance.

It is clear that the research works above, although a sample, provide an extensive analysis of data assetization from multiple perspectives. However, there are not guidelines allowing a successful data assetization journey in terms of identifying data eligible for assetization, making assetization progress, suspending/stopping assetization, governing data as an asset, etc. In the rest of this paper, we present our data assetization journey.

2.2 Case Study

To illustrate data that could be part of an assetization journey, we adopt the case study of a small-size financial company that processes credit applications on behalf of a group of large commercial banks [14]. Figure 1 is the BPMN-based credit application's process model consisting of activities (e.g., *check-for-completeness* and *request-credit-card*), gateways (e.g., *application-completeness* and *credit-amount*), and flows. When an online credit application is received, a staff verifies the application. Should the application be incomplete, the staff will ask the applicant for the missing documents. Otherwise, more checks take place with the support of third parties like the central bank depending on the credit's amount (e.g., up to $500). Then, the staff sends the on-duty manager the application for final decision. If accepted, the applicant will be notified and a credit-card production request will be issued. Otherwise, the applicant will be informed of the rejection.

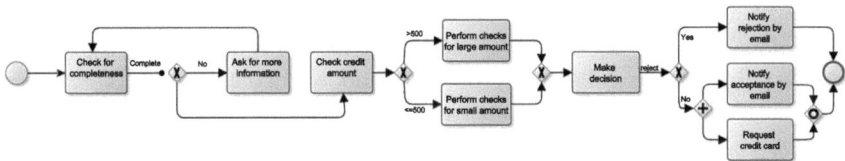

Fig. 1. Process model of the credit application BP ([14]).

Multiple ODRL policies could be associated with the data inherent to the credit application BP. By converting data into assets, policies would control the actions that are permitted to exercise over data, that are prohibited from exercising over data, and that must be exercised over data. For illustration, in Listing 1.1, an agreement policy refers to the assigner central-bank (line 7), the assignee financial-company-staff (line 8), the asset customer_ssn (line 6), and, finally, the obligation rule (line 5) to have the action modify (line 10) exercised over this asset from a specific location (lines 11–14).

Listing 1.1. ODRL policy over data

```
{
    "@context": "http://www.w3.org/ns/odrl.jsonld",
    "uid": "http://example.com/policy:101",
    "@type": "Agreement",
    "obligation": [{
        "target": "http://example.com/asset/customer_ssn",
        "assigner": "http://example.com/CentralBankDBOwner:org:CBDB",
        "assignee": "http://example.com/staff/FinancialBank:org:FB",
        "action": [{
            "rdf:value": {"@id":"odrl:modify"},
            "refinement": [{
                "leftOperand": "spatial",
                "operator": "eq",
                "rightOperand": "X"
            }]
        }]
    }]
}
```

3 Assetization Journey

The completion of the data assetization journey is dependent on defining the intrinsic features of both data and asset. Steps, techniques, and modules performing the journey and thus, acting on these features also need to be defined.

3.1 Intrinsic Features of Data

The following intrinsic features characterize data from multiple perspectives allowing to set-up what we could refer to as a **data quality-model**.

- Freshness (timeliness) is how critical data must be up-to-date/recent since the last time it was collected, processed, and/or distributed.
- Integrity is how critical data must remain consistent from its collection/generation to disposal.
- Availability is how critical data must be either collected or generated for use, and for how long data remains available before disposal.
- Confidentiality is how critical data must be kept private before it becomes public.
- Shareability is how critical data must be made available for multiple parties.
- Usefulness is how critical data must achieve fitness-for-purpose.

The features above aim at preparing data to become assets. Indeed, a data assetization journey would assess the impact of each feature on an organization's ongoing practices. Along with these features, a report by Accenture AWS Business Group identifies 3 types of value created with data: transactional value allowing to understand and execute business transactions, informational value allowing to describe past performance and draw conclusions, and analytical value allowing to automate activities, guide decisions, and predict outcomes [1].

3.2 Intrinsic Features of Assets

The following intrinsic features characterize asset from multiple perspectives allowing to set-up what we could refer to as an asset operation-model.

- Depreciation is about the (selling) value and/or (ongoing) performance of assets over time. In data assetization, inappropriate or limited handling of organizations' structural and functional changes could make data as an asset depreciate in the sense of becoming irrelevant or useless.
- Transferability is about the change of assets' ownerships and/or locations with respect to legal regulations of these ownerships/locations. In data assetization, transferability permits to exchange data as an asset internally and externally while satisfying requirements like cross-border and privacy.
- Disposability[3] is about the unavailability of assets for use due to for instance, expiry date and being worn out, which enforces the depreciation feature. In data assetization, disposability allows releasing unnecessary data as an asset with the option of generating new data as future assets, if deemed necessary.
- Convertibility is about the monetization of assets so, they become cash. In data assetization, data as an asset could be sold to third parties.

The features above allow to track the evolution of data after becoming assets. In Sect. 1, we mention the limited expressiveness of DaaS to capture this evolution. In real life, data depreciates, is transferred, is disposed of, and/or is monetized. By analogy with the asset's features above, Birch and Muniesa expose an asset's dimensions in terms of temporality (an asset has to have a useful life beyond one year), separability (an asset has to be distinguishable from an entity), economic control (an entity has to be able to assert a claim to its economic benefits), and future benefits (an entity has to be able to secure future economic benefits) [5].

3.3 Overview of the Journey

Figure 2 illustrates our proposed data assetization journey having data as a starting point, asset as an ending point, and eligibility, enrichment, and governance as 3 stopovers between these 2 points. Our journey is quite different from the journey of Wang who refers to data collection, storage, processing, analysis, and application [32]. While Wang stresses the 4 elements that should characterize data assets namely, assets should originate from past production and operation activities, ownership relationship of the assets is clear, assets can bring economic benefits to the owner, and assets exist in electronic form, we, at our end, stress the chronology of completing the assetization journey based on stopovers where particular operations are carried out and particular techniques empower these operations. Our journey is also quite different from Xu et al.'s data assetization framework who refer to human/technology/data resource preparation, dynamic/network capability construction, and value realization in terms of process

---

[3] Some refer to perishable.

optimization, product development, and business innovation [34]. We focus on data in terms of eligibility/which data become assets, enrichment/how to prepare data for assetization, and governance/how to manage data as an asset. Finally, a potential data assetization journey by Accenture AWS Business Group suggests provenance, curation, and utilization principles to help organizations transform data into a digital working asset [1]. These principles manage tangible assets, such as tracking the source of fresh produce or the location of a product's manufacturer.

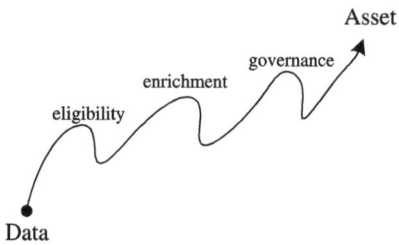

Fig. 2. Stopovers inherent to the data assetization journey.

3.4 Eligibility Stopover

It ensures that data is "worth" assetizing; not all data can become assets. In this case, data is a good [19] and is not inherently asset-oriented [31]. We analyze eligibility of data for assetization from 3 perspectives: finance, legal, and technology.

- From a finance perspective, the focus in on establishing the cost of generating, collecting, safeguarding, processing, storing, distributing, maintaining, and disposing of data. To what extent is an organization willing to spend on data assetization on top of this cost? We expect that assetization would incur additional costs due to operations related to monitoring assets, monetizing assets, restricting assets, etc. To value data based on how it is used and by whom[4], 3 methods could be used [19]:
  1. Cost-based methods deal with how data is collected, stored, and analyzed.
  2. Market-based methods deal with market prices of data or market valuations of companies that use data intensively.
  3. Use-based methods deal with how data is used, in what context, and for what purpose. These methods also consider consumers' willingness to pay for data and businesses' expected profits.
- From a legal perspective, the focus is on maintaining data quality and privacy as well as identifying data sources. To what extent is an organization willing to "comprise" this quality and privacy and to reveal these sources for the needs of assetization? We expect that assetization would raise a new

---

[4] How to price a data asset, pivotal.substack.com/p/how-to-price-a-data-asset.

set of legal requirements different from those related to data. Here, Machine Learning (ML) could be invaluable for detecting patterns, anomalies, and inconsistencies, so that data quality is improved by identifying and fixing errors. Techniques like anomaly detection [17] can flag outliers, while supervised learning models trained on high-quality datasets can automate data cleaning, reducing manual errors and efforts. This guarantees that data is not only legally compliant but also reliable enough to be treated as an asset. Generative AI could offer unique support, especially in data-limited scenarios where comprehensive datasets are needed. For instance, generative models, like Generative Adversarial Networks (GANs) [11], can create synthetic data to supplement existing datasets. This synthetic data fills gaps, making data assets more comprehensive and valuable for analytics and AI applications. Additionally, some AI techniques enable privacy-preserving data sharing. Federated learning for instance, supports collaborative ML across multiple organizations without sharing raw data, preserving privacy while enabling shared insights from distributed datasets, thus, expanding the potential value of data assets within regulatory boundaries [28].

- From a technology perspective, the focus is on identifying technical means for managing data. To what extent is an organization willing to reconfigure these means for the needs of assetization? We expect that assetization would raise a new set of technical requirements different from those related to data. For instance, data integration tools, which traditionally support the merging of data from different sources, may need to become more sophisticated to ensure data interoperability across departments and even organizations. Data lakes and data warehouses are popular options for centralized data storage, but in an assetization framework, they may need enhanced capabilities to support metadata indexing, version control, and retrieval speed. Blockchain technology could also play a significant role in data traceability and ownership, especially in multi-party settings where data is treated as a valuable asset. Using blockchain to create an immutable record of data transactions and usage can provide transparency, allowing stakeholders to track how data assets are accessed and utilized over time. Last but not least, automation technologies such as Robotic Process Automation (RPA) and Automated Machine Learning (AutoML) can streamline data preparation, transformation, and labeling tasks, reducing human errors and operational costs. RPA bots can automate repetitive data management tasks, like data extraction, cleaning, or updating, ensuring that data assets are kept up-to-date and compliant with assetization standards. Similarly, AutoML tools, by automating model selection and tuning, enable non-technical users to leverage ML, further enhancing the usability and accessibility of data as an asset.

By analyzing data assetization from finance, legal, and technology perspectives, organizations can determine the eligibility of data for assetization, making it ready to deliver a real strategic value to their owners. Figure 3 illustrates the trade-off between eligibility of assetization and efforts of assetization using a scale of low *versus* high. The efforts refer to what assetization would trigger as per each perspective discussed above. The figure also identifies 4 cases related to data assetization in terms of highly recommended due to high eligibility/low efforts, recommended due to high eligibility/high efforts, not recommended due to low eligibility/low efforts, and not recommended due to low eligibility/high efforts.

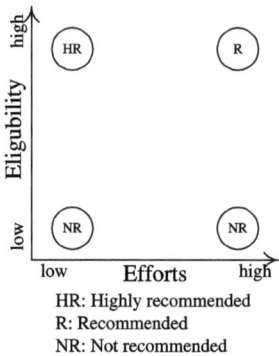

Fig. 3. Trade-off between eligibility and efforts of assetization.

3.5 Enrichment Stopover

It provides additional details (like meta-data) about data, which should permit to "reason" over these data after passing the eligibility stopover successfully. These details are about the intrinsic features of data as per Sect. 3.1 where values of data, requirements on data, and/or constraints on data are set. It is stated that *"data enrichment is the process of enhancing raw data by merging it with additional information from various sources to improve its accuracy and completeness"*[5]. To action the enrichment stopover that must be compliant with any existing data laws and regulations such as General Data Protection Regulation (**GDPR**) and California Consumer Privacy Act (**CCPA**), we adopt 2 examples of data, *customer_ssn* and *credit_rate*, from the case study. Table 1 illustrates how we enrich data based on their nature where some intrinsic features are deemed not relevant while others shed light on requirements like remaining private until disposal. Each feature contributes differently to the **governance** stopover as it will be discussed later. This stopover's policies will enforce the role of data's features in the assetization journey.

An important aspect of data enrichment is metadata management, a cornerstone of data assetization. Using Natural Language Processing (**NLP**) techniques [33], metadata for unstructured data such as text documents, emails, or customer feedback can be automatically generated and organized. NLP models like Latent Dirichlet Allocation [6] and Bidirectional Encoder Representations from Transformers (**BERT**) [26] can extract keywords, categorize information,

---
[5] www.fullstory.com/blog/data-enrichment.

Table 1. Examples of data enrichment based on features

Data	Feature	Description
customer_ssn	Freshness	Not relevant
	Integrity	Must be maintained all the time
	Availability	Must exist for customer identification
	Confidentiality	Must remain private all the time
	Shareability	Must not be updated concurrently
	Usefulness	Must exist for customer identification
credit_rate	Freshness	Must be updated regularly
	Integrity	Must be maintained all the time
	Availability	Must exist for installment calculation
	Confidentiality	Must be made public
	Shareability	Must not be updated concurrently
	Usefulness	Must exist for installment calculation

and even summarize text, simplifying classification and search within datasets. Effective metadata management enhances data discovery and usability, so that the right information is available when needed, thereby maximizing data value as an asset. Another important aspect of data enrichment is labeling new data, a key application of data valuation. Due to labeling costs, it is often necessary to carefully select which data to label next. Here, active learning algorithms (e.g., [29]) enhance data labeling by selecting groups of new data samples that optimize model learning performance.

3.6 Governance Stopover

It develops policies to manage data as an asset in terms of depreciation, transferability, disposability, and convertibility. For each asset's intrinsic feature, policies will be defined covering data's intrinsic features set in the enrichment stopover.

– Depreciation policies preserve the value of data over time by implementing strategies that would mitigate their immediate degradation in terms of quality, relevance, or usability [3]. Organizations often implement measures such as regular quality assessment, data validation, and scheduled updates, so that datasets remain actionable. For instance, automated tools can monitor data integrity, flagging obsolete or inaccurate records for archival or removal. Archiving rules may also be defined to preserve the historical value of data, such as maintaining customer transaction records for trend analysis. By proactively managing depreciation, organizations can ensure that their data retain its utility and relevance.
Illustration: a depreciation policy covers data **freshness**, e.g., *credit_rate* (Table 1), so that these data remain up-to-date through regular updates. This

is represented with Listing 1.2 where a request policy (line 3) has an obligation rule (line 6) to have the financial company (line 8) update the *credit_rate* (line 7) on a daily basis using the action modify (line 10).

Listing 1.2. ODRL depreciation policy covering freshness

```
 1  {
 2      "@context": "http://www.w3.org/ns/odrl.jsonld",
 3      "@type": "Request",
 4      "uid": "http://example.com/policy:001",
 5      "profile": "http://example.com/odrl:profile:01",
 6      "obligation": [{
 7          "target": "http://example.com/data/credit_app/credit_rate",
 8          "assignee": "http://example.com/staff/FinancialBank:org:FC",
 9          "action": [{
10              "rdf:value": {"@id":"odrl:modify"},
11              "refinement": [{
12                  "leftOperand": "elapsedTime",
13                  "operator": "lteq",
14                  "rightOperand":{"@value":"P24H", "@type":"xsd:duration"
                      },
15                  "comment":"less than 24 hours"
16              }]
17          }]
18      }]
19  }
```

- Transferability policies will govern the secure and compliant sharing of data across different owners, platforms, departments, organizations, and/or locations [22]. These policies would often depend on secure transfer protocols such as Hyper Text Transfer Protocol Secure (HTTPS) and Secure File Transfer Protocol (SFTP), encryption to protect data during transmission, and role-based access systems to ensure only authorized individuals have access. Additionally, compliance with laws such as GDPR and Health Insurance Portability and Accountability Act (HIPAA) may require data anonymization or formal agreements outlining usage rights and responsibilities [16]. For instance, a company sharing operational data with external partners can enforce encryption and contractual obligations to safeguard proprietary information [21].

Illustration: a transferability policy covers data confidentiality, e.g., *customer_ssn* (Table 1), ensuring that these data comply with the new location's regulations after their transfer. This is represented with Listing 1.3 where a request policy (line 3) has an obligation rule (line 5) to have the central bank (line 7), upon receiving a *customer_ssn* (line 6) for auditing purposes (lines 11–14), maintain the confidentiality of this data during the audit (lines 8–9).

Listing 1.3. ODRL transferability policy covering confidentiality

```
1  {
2      "@context": "http://www.w3.org/ns/odrl.jsonld",
3      "@type": "Request",
4      "uid": "http://example.com/policy:002",
5      "obligation": [{
6          "target": "http://example.com/deature/customer_ssn",
7          "assignee": "http://example.com/staff/FinancialBank:org:FC",
8          "action": "reviewPolicy",
9          "uid": "http://example.com/policy:111",
```

```
10          "constraint": [{
11              "leftOperand": "event",
12              "operator": "eq",
13              "rightOperand":"https://www.wikidata.org/wiki/Q2918584",
14              "comment": "relocation"
15          }]
16      }]
17  }
```

- Disposability policies will consider the secure and lawful removal of data that is no longer needed or has reached the end of its lifecycle. These policies typically include retention schedules that define how long data should be kept, deletion standards that specify techniques for destroying data, and alignment with regulatory requirements such as GDPR's "Right to Erasure". Secure disposal practices, such as cryptographic wiping and physical destruction of records, reduce the risk of data breaches. For instance, a financial institution might delete customer financial records after a defined period unless explicitly required for legal investigations, allowing compliance while optimizing storage resources [24].

Illustration: a disposability policy covers data availability, e.g., $customer\_slr$ (salary), allowing the removal of data that is no longer intended for a certain purpose. This is represented with Listing 1.4 where a request policy (line 3) has an obligation rule (line 5) to have the financial bank (line 7) delete (line 8) the $customer\_slr$ when customers' creditworthiness based on incomes is no longer accepted for loan renewal (lines 10–13).

Listing 1.4. ODRL disposability policy covering availability

```
1   {
2       "@context": "http://www.w3.org/ns/odrl.jsonld",
3       "@type": "Request",
4       "uid": "http://example.com/policy:003",
5       "obligation": [{
6           "target": "http://example.com/deature/customer_slr",
7           "assignee": "http://example.com/staff/FinancialBank:org:FC",
8           "action": "delete",
9           "constraint": [{
10              "leftOperand": "context",
11              "operator": "eq",
12              "rightOperand": "https://www.wikidata.org/wiki/Q65722561",
13              "comment": "loan renewal"
14          }]
15      }]
16  }
```

- Convertibility policies will regulate the process of monetizing data, transforming it into a revenue-generating asset [23]. These policies will be guidelines for assessing and maximizing the financial value of data while maintaining compliance with legal and ethical standards. To this end, they may include data valuation methods like those listed in the eligibility stopover. Along with these methods, a study in [18] discusses multiple ways of valuing data as an asset, highlighting the importance of accurate valuation in data monetization strategies. Convertibility policies may also focus on packaging data into marketable products or services (referred to as data productization), such as providing customer behaviors' insights to advertisers or brands. The work of

Thota et al. in [27] emphasized the significance of developing data products that align with customer needs and business objectives to effectively monetize data [35]. Finally, organizations may also participate in data marketplaces, where governance focuses on secure and compliant sharing of data through encryption, anonymization, or differential privacy techniques. A review in [23] explores the challenges and opportunities of data trading and monetization, underscoring the need for robust governance framework.

Illustration: a convertibility policy covers data shareability, e.g., *customer_prf* (profile), by "selling" this data to third parties. This is represented with Listing 1.5 where an agreement policy (line 3) between the financial bank (line 8) and a third party (line 7) has a permission rule (line 5) to have the financial bank share (line 9) the *customer_prf* (line 6) in return of compensation (lines 12–16).

Listing 1.5. ODRL convertibility policy covering shareability

```
 1  {
 2      "@context": "http://www.w3.org/ns/odrl.jsonld",
 3      "@type": "Agreement",
 4      "uid": "http://example.com/policy:004",
 5      "permission": [{
 6          "target": "http://example.com/deature/customer_prf",
 7          "assigner": "http://example.com/staff/FinancialBank:org:FC",
 8          "assignee": "http://example.com/staff/FinancialBankThirdParty:
               TP",
 9          "action": "shareAlike",
10          "duty": [{
11              "action": [{
12                  "rdf:value": {"@id":"odrl:compensate"},
13                  "refinement": [{
14                      "leftOperand": "payAmount",
15                      "operator": "gteq",
16                      "rightOperand": "https://www.wikidata.org/wiki/Q179
                           222"
17                  }]
18              }]
19          }]
20      }]
21  }
```

4 Experiments

To demonstrate the technical feasibility of the data assetization journey, we developed a prototype with focus on the eligibility stopover. This prototype, referred to as *Data Eligibility Assessment*, assists organizations assess and classify datasets based on their eligibility as potential valuable assets.

In terms of technologies, we used Streamlit, a Python-based Web application framework for building interactive Web interfaces. It employs Pandas for dataset manipulation and preprocessing, and NumPy for efficient numerical computations during score calculation and normalization. To provide recommendations for eligibility assessment, we also used OpenAI's GPT models via the OpenAI

API, leveraging few-shot learning techniques to generate context-specific suggestions for feature weighting and eligibility threshold setting. The architecture of the prototype is designed to be extensible, with LangChain framework used to enable integration with alternative large language models, thus supporting future customization and model evolution. Users can interactively upload datasets, define business domains and assessment contexts, adjust feature weights, set a user-defined eligibility threshold, and generate normalized eligibility scores. The threshold setting allows users to flexibly determine the minimum eligibility score a dataset must achieve to be considered an asset, providing greater control over assetization criteria based on organizational needs and priorities. The workflow of operations spans over 6 steps:

1. *Dataset upload & visualization*: users upload the dataset to assess as a CSV file. They also have the option of visualizing its content, so that available numeric features are identified (Fig. 4).
2. *Domain and context specification*: users select a business domain (e.g., logistics, finance, and healthcare), so that a contextual analysis of the dataset is set according to their specific objectives.
3. *AI-based recommendations*: GPT models are used to recommend feature weights and an eligibility threshold. A justification for the recommendations is also provided (Fig. 5).
4. *Customization of weights and thresholds*: users can manually adjust the recommended feature weights (ranging from −1.0 to +1.0) and eligibility threshold (normalized between 0 and 100) to better reflect their needs (Fig. 6).
5. *Eligibility score calculation*: raw and normalized eligibility scores (Eqs. 1 and 2.) are computed for each record based on the customized configuration.
6. *Result visualization and export*: finally, users can view a summary of eligible *versus* ineligible records and download the detailed results as a CSV file (Fig. 7).

Score calculation. For each record i, the raw eligibility score is computed as:

$$raw\_eligibility\_score_i = \sum_{j=1}^{N} (x_{ij} \times w_j) \qquad (1)$$

where:

- x_{ij} is the value of feature j for record i,
- w_j is the weight assigned to feature j (between −1.0 and 1.0),
- And, N is the number of selected features.

After computing the raw scores for all records, the normalized eligibility score for each record i is calculated as:

$$normalized\_eligibility\_score_i = \left(\frac{raw\_score_i - min\_score}{max\_score - min\_score} \right) \times 100 \qquad (2)$$

where min_score and max_score represent the minimum and maximum raw scores across all records, respectively. This normalized scoring approach ensures that eligibility assessments remain consistent across diverse datasets and contexts.

The *Data Eligibility Assessment* prototype primarily operationalizes the eligibility dimension presented in Fig. 3. By providing normalized eligibility scores based on configurable criteria and AI-driven recommendations, we enable organizations to assess which datasets are highly promising for assetization. While the estimation of efforts (i.e., financial, legal, and technological costs) is not automated in the current version, the eligibility scores support decision-makers in positioning their datasets within the eligibility-effort trade-off space. This quantification of eligibility facilitates a more informed prioritization of datasets for assetization initiatives.

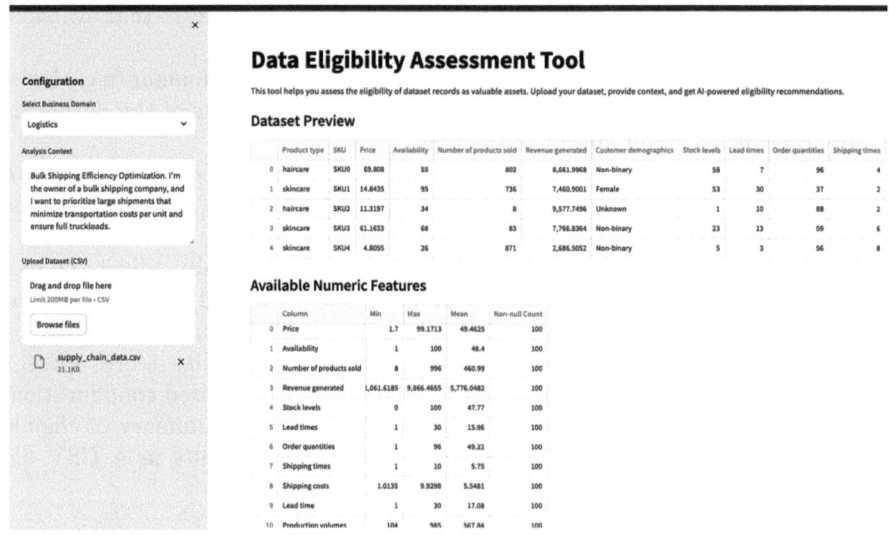

Fig. 4. Dataset upload, preview, and numeric feature identification.

5 Discussions

In this paper, we designed a "guided" journey to transform data into organizational assets through a structured, policy-driven way of doing. We highlighted the need for a systematic process to assess, enrich, and govern data, so that economic value and operational utility are established for decision makers. To respond to this need, 3 stopovers are part of the journey: eligibility, enrichment, and governance covering each specific aspects of the assetization journey. The eligibility stopover determines whether data qualifies to become an asset, the enrichment stopover enhances data with metadata and additional information to improve its quality and utility, and the governance stopover establishes policies for managing

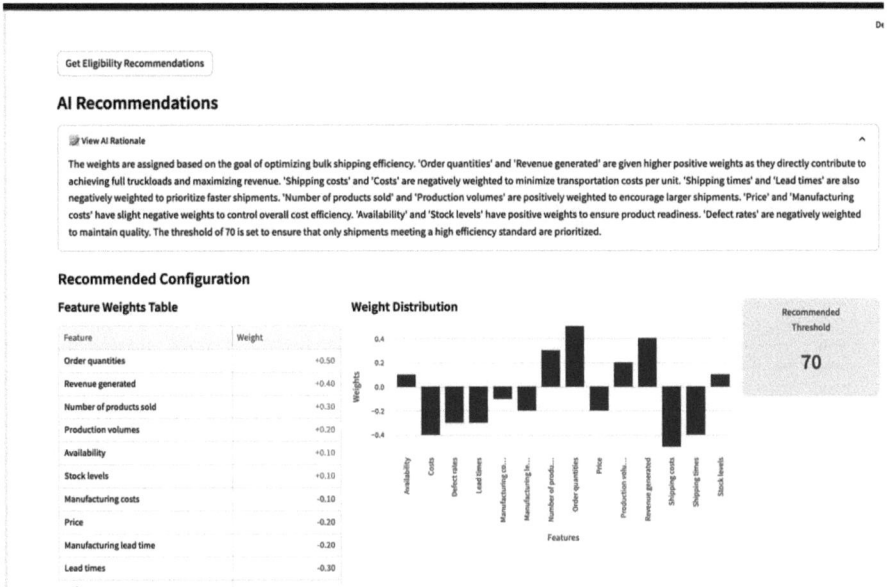

Fig. 5. AI-powered feature weight recommendations and threshold suggestion.

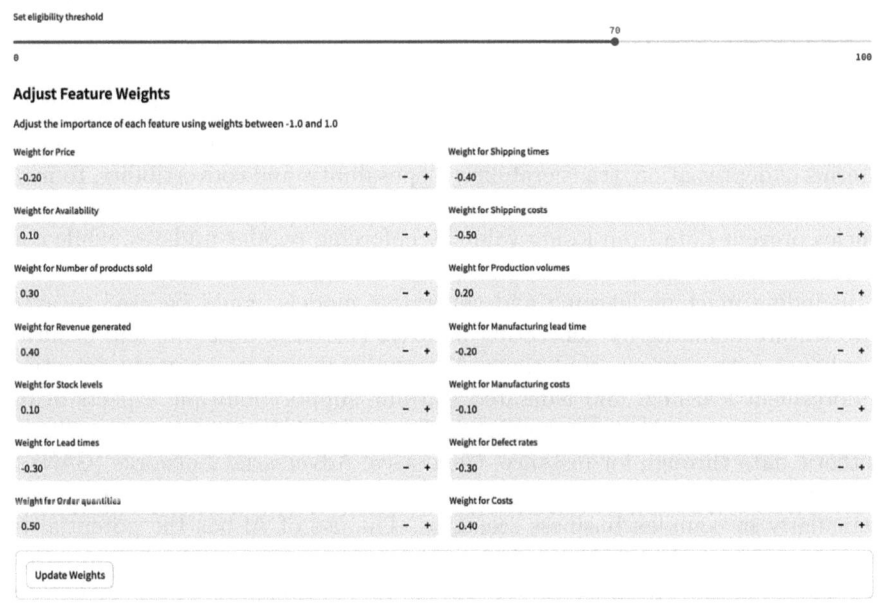

Fig. 6. Manual adjustment of feature weights and eligibility threshold.

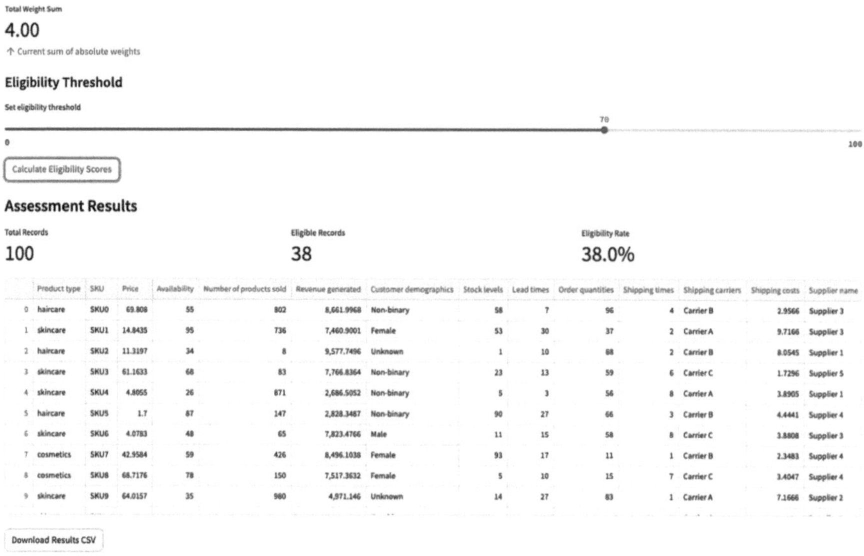

	Product type	SKU	Price	Availability	Number of products sold	Revenue generated	Customer demographics	Stock levels	Lead times	Order quantities	Shipping times	Shipping carriers	Shipping costs	Supplier name
0	haircare	SKU0	69.808	55	802	8,661.9968	Non-binary	58	7	96	4	Carrier B	2.9566	Supplier 3
1	skincare	SKU1	14.8435	95	736	7,460.9001	Female	53	30	37	2	Carrier A	9.7166	Supplier 3
2	haircare	SKU2	11.3197	34	8	9,577.7496	Unknown	1	10	88	2	Carrier B	8.0545	Supplier 1
3	skincare	SKU3	61.1633	68	83	7,766.8364	Non-binary	23	13	59	6	Carrier C	1.7296	Supplier 5
4	skincare	SKU4	4.8055	26	871	2,686.5052	Non-binary	5	3	56	8	Carrier A	3.8905	Supplier 1
5	haircare	SKU5	1.7	87	147	2,828.3487	Non-binary	90	27	66	3	Carrier B	4.4441	Supplier 4
6	skincare	SKU6	4.0783	48	65	7,823.4766	Male	11	15	58	8	Carrier C	3.8808	Supplier 3
7	cosmetics	SKU7	42.9584	59	426	8,496.1038	Female	93	17	11	1	Carrier B	2.3483	Supplier 4
8	cosmetics	SKU8	68.7176	78	150	7,517.3632	Female	5	10	15	7	Carrier C	3.4047	Supplier 4
9	skincare	SKU9	64.0157	35	980	4,971.146	Unknown	14	27	83	1	Carrier A	7.1666	Supplier 2

Fig. 7. Eligibility score calculation and assessment results preview.

data as an asset over its lifecycle. We recommended supporting these stopovers with AI techniques and formalizing them through ODRL policies, which enforce permissions, prohibitions, and obligations that steer the assetization journey's progress.

An essential aspect of our work is the focus on the intrinsic features of both data and assets. We evaluated data based on freshness, integrity, availability, confidentiality, shareability, and usefulness features, so that data is properly assessed before being considered an asset. Similarly, we outlined asset features in terms of depreciation, transferability, disposability, and convertibility, to manage data assets effectively over time through policies. For instance, depreciation policies prevent data from losing value by enforcing regular updates, while convertibility policies guide data monetization strategies. We also discussed how AI, including machine learning and generative models, could be used to automate feature handling by generating policies, enriching datasets, and ensuring compliance with privacy regulations. AI techniques such as anomaly detection, reinforcement learning, and federated learning support multiple aspects of the assetization journey. AI automates data validation and cleaning processes, creates synthetic data through for instance, Generative Adversarial Networks (GAN) to handle incomplete datasets, and dynamically generates policies for unseen cases, particularly in complex business scenarios. The use of AI has the potential to, not only, enhance the efficiency of the assetization journey but, also, to ensure adaptability to new and unforeseen requirements, making it a scalable and robust approach to manage data assets.

The case study in this paper also permitted to demonstrate the feasibility of the assetization journey where a financial institution's credit application process was adopted. The journey illustrates how ODRL policies can be applied to manage sensitive data, such as customer social security numbers and credit rates, across various scenarios. Policies also enforce obligations, such as updating credit rates daily to maintain freshness, and define permissions for secure data sharing while preserving confidentiality. These examples effectively showcase how we can handle real-world data by adopting AI and policies.

Despite the assetization journey's strengths, some challenges lay ahead. One challenge is scalability, particularly in handling large datasets and highly fragmented processes. While AI techniques improve efficiency, their computational requirements and scalability in real-time systems remain areas for further research. Another challenge is the potential of conflicts between ODRL policies, especially when multiple fragments of a process have inter-dependencies. Additionally, addressing legal and ethical concerns associated with cross-border data transfer, data ownership, and evolving regulations, represents another important challenge in this work.

Finally, this work makes a solid foundation on which future research can be built by focusing on dynamic and adaptive policy generation using advanced AI techniques such as deep reinforcement learning and graph neural networks. These techniques can address unseen scenarios, particularly in managing highly fragmented processes. Another important direction is enhancing policy transparency through eXplainable AI (XAI) models, so that AI-generated policies are interpretable and auditable for regulatory compliance. Blockchain technology could also be adopted to provide immutable logging and auditing of policy enforcement, improving security and accountability. Additionally, future work could explore refining economic valuation models for data assets, incorporating factors such as usage frequency, predictive potential, and quality to develop more accurate assessments of data's financial worth. Lastly, deploying and evaluating the assetization journey in real-world applications across industries, such as healthcare, manufacturing, and finance, would provide insights into its practical implementation and scalability challenges.

6 Conclusion

This paper presented a data assetization journey that aims at enabling a smooth and successful transition of data into assets. The journey builds upon the intrinsic features of both data and asset to put forward a set of operations to perform at specific stopovers referred to as eligibility, enrichment, and governance. In the first stopover, data is analyzed from finance, legal, and technology perspectives to decide on its appropriateness for assetization. This appropriateness was demonstrated through an in-house prototype that for the moment targets the first stopover where datasets eligibel for assetization are identified. Passing this first stopover successfully, the second stopover provides additional details about data to identify potential values of data, potential requirements on data, and potential constraints on data. Finally, the third stopover focuses on developing policies in ODRL to manage data evolution in terms of depreciation, transferability, disposability, and convertibility. The journey's stopovers were also complemented with different AI techniques permitting to handle unseen assetization cases, learn from past assetization cases, and generate policies. On top of applying the data assetization journey to real-world scenarios, our future work includes scalability when large datasets are subject to assetization and policy consistency to avoid conflicts.

Disclosure of Interests. The authors have no competing interests to declare that are relevant to the content of this article.

References

1. Axson, D., Escapa, C.: The CFO's guide to data assetization and monetization. d1.awsstatic.com/executive-insights/en_US/ebook-accenture-cfo-guide-to-data-assetization-and-monetization.pdf (2024 (2024))
2. Beauvisage, T., Mellet, K.: Turning Things into Assets. chap. Datassets: Assetizing and Marketizing Personal Data (2020)
3. Bernardi, F.A., Alves, D., Crepaldi, N., Yamada, D.B., Lima, V.C., Rijo, R.: Data quality in health research: integrative literature review. J. Med. Internet Res. **25** (2023)
4. Birch, K., Komljenovic, J., Sellar, S., Hansen, M.: Data as Asset, Data as Rent? Rentiership Practices in EdTech Startups. Learning, Media and Technology (2024)
5. Birch, K., Muniesa, F.: Assetization. Turning Things into Assets in Technoscientific Capitalism. The MIT Press, Cambridge, Mass (2020)
6. Blei, D., Ng, A., Jordan, M.: Latent Dirichlet allocation. J. Mach. Learn. Res. **3** (2003)
7. Cano-Benito, J., Cimmino, A., García-Castro, R.: Injecting data into ODRL privacy policies dynamically with RDF mappings. In: Companion Proceedings of the ACM Web Conference 2023 (WWW'2023). Austin, TX, USA (2023)
8. Carrascal, J., Riederer, C., Erramilli, V., Cherubini, M., de Oliveira, R.: Your browsing behavior for a big Mac: economics of personal information online. In: Proceedings of the 22nd International Conference on World Wide Web (WWW'2013). Rio de Janeiro, Brazil (2013)
9. Chen, L.: Data assetization and capital market information efficiency: evidence from hidden champion SMEs in China. Fut. Bus. J. **10**(110) (2024)
10. Coyle, D., Diepeveen, S., Wdowin, J., Kay, L., Tennison, J.: The Value of Data. Tech. rep., Bennett Institute for Public Policy, Cambridge, UK. https://www.bennettinstitute.cam.ac.uk/wp-content/uploads/2020/12/Value_of_data_summary_report_26_Feb.pdf (2020)
11. Faisal, R., Claudio, S., Sergio, C., , Diego, R.: Generative adversarial networks for synthetic data generation in finance: evaluating statistical similarities and quality assessment. AI **5**(2) (2024)
12. Hackfort, S., Marquis, S., Bronson, K.: Harvesting value: corporate strategies of data assetization in agriculture and their socio-ecological implications. Big Data Soc. **11**(1) (2024)
13. Huberman, B., Adar, E., Fine, L.: Valuating privacy. IEEE Secur. Priv. **5**(3) (2005)
14. Kallel, S., et al.: Restriction-based fragmentation of business processes over the cloud. Concur. Comput. Pract. Exper. **37** (2021)

15. Kang, C., Yang, T., Yan, W.: Research on the Dilemma of data assetization and solutions. In: 2024 International Conference on Big Data and Digital Management (ICBDDM'2024). Shanghai, China (2024)
16. Kevin, M.: Data Privacy and Data Protection Issues in Cloud Computing. Cambridge University Press (2021)
17. Koren, O., Koren, M., Peretz, O.: A procedure for anomaly detection and analysis. Eng. Appl. Artif. Intell. **117** (2023)
18. Laura, V.: Valuing data as an asset. Rev. Finan. **27**(5) (01 2023)
19. Ltd, F.E.: The Value of Data Assets - A Report for the Department for Digital, Culture, Media and Sport. Tech. rep., Frontier Economics Ltd (2021)
20. Maamar, Z., Benna, A., Chikhaoui, B.: Data as an Asset: Challenges and Research Directions. IEEE IT Professional (2025 (forthcoming))
21. Marcucci, S., Alarcon, N., Verhulst, S., Wullhorst, E.: Informing the global data future: benchmarking data governance frameworks. Data Pol. **5** (2023)
22. Mistale, T.: Enabling Transatlantic Trade and Protecting Privacy through Cross-Border Data Transfer Agreements. Cambridge University Press (2023)
23. Ofulue, J., Benyoucef, M.: Data monetization: insights from a technology-enabled literature review and research agenda. Manag. Rev. Quart. **74**(2) (2024)
24. Sadowski, J.: When data is capital: datafication, accumulation, and extraction. Big Data Soc. **6**(1) (2019)
25. Srivastava, G., Flath, C.M., Lin, J.C., Zhang, Y.: Challenges and outcomes using big data as a service. Bus. Info. Syst. Eng. **1**(66) (2024)
26. Sun, C., Qiu, X., Xu, Y., Huang, X.: How to fine-tune BERT for text classification? Tech. rep., arXiv, https://arxiv.org/abs/1905.05583 (2020)
27. Thota, B., Swartz, J., Kuchembuck, R., Gandhi, S.: Demystifying Data Monetization. MIT Sloan Management Review (2018)
28. Truong, N., Sun, K., Wang, S., Guitton, F., Guo, Y.: Privacy preservation in federated learning: an insightful survey from the GDPR perspective. Comput. Secur. **110** (2021)
29. Ul Amin, S., Hussain, A., Kim, B., Seo, S.: Deep learning based active learning technique for data annotation and improve the overall performance of classification models. Expert Syst. Appl. **228** (2023)
30. W3C: ODRL Information Model 2.2. https://www.w3.org/TR/2018/REC-odrl-model-20180215/ (2018). (Visited in January 2022)
31. Wang, J., Li, Y., Song, W., Li, A.: Research on the theory and method of grid data asset management. In: Proceedings of the 6th International Conference on Information Technology and Quantitative Management (ITQM'2018). Omaha, Nebraska, USA (2018)
32. Wang, X.: Research on data assetization difficulties and management path. Front. Bus. Econom. Manag. **11**(1) (2023)
33. Waterworth, D., Sethuvenkatraman, S., Sheng, Q.: Advancing smart building readiness: automated metadata extraction using neural language processing methods. Adv. Appl. Energ. **3** (2021)
34. Xu, T., Shi, H., Shi, Y., You, J.: From data to data asset: conceptual evolution and strategic imperatives in the digital economy era. Asia Pacif. J. Innov. Entrepr. **18**(1) (2023)
35. Zhuo, W., et al.: Research on productization and development trend of data desensitization technology. In: Proceedings of the IEEE 20th International Conference on Trust, Security and Privacy in Computing and Communications (TrustCom'2021). Shenyang, Republic of China (2021)

Do Echo Top Heights Improve Deep Learning Rainfall Nowcasts? A Case Study in the Netherlands

Peter Pavlík[1,2(✉)], Marc Schleiss[3], Anna Bou Ezzeddine[1], and Viera Rozinajová[1]

[1] Kempelen Institute of Intelligent Technologies, Bratislava, Slovakia
peter.pavlik@kinit.sk
[2] Faculty of Information Technology, Brno University of Technology, Brno, Czech Republic
[3] Department of Geoscience and Remote Sensing, Delft University of Technology, Delft, The Netherlands

Abstract. Precipitation nowcasting – the short-term prediction of rainfall using recent radar observations – is critical for weather-sensitive sectors such as transportation, agriculture, and disaster mitigation. While recent deep learning models have shown promise in improving nowcasting skill, most approaches rely solely on 2D radar reflectivity fields, discarding valuable vertical information available in the full 3D radar volume. In this work, we explore the use of echo top height (ETH), a 2D projection indicating the maximum altitude of radar reflectivity above a given threshold, as an auxiliary input variable for deep learning-based nowcasting. We examine the relationship between ETH and radar reflectivity, confirming its relevance for predicting rainfall intensity. We implement a single-pass 3D U-Net that processes both the radar reflectivity and ETH as separate input channels. While our models are able to leverage ETH to improve skill at low rain-rate thresholds, results are inconsistent at higher intensities and the models with ETH systematically underestimate precipitation intensity. Three case studies are used to illustrate how ETH can help in some cases, but also confuse the models and increase the error variance. Nonetheless, the study serves as a foundation for critically assessing the potential contribution of additional variables to nowcasting performance.

Keywords: Precipitation Nowcasting · Deep Learning · Weather Radar

1 Introduction

Nowcasting is defined by the World Meteorological Agency as forecasting with local detail, by any method, over a period from the present to six hours ahead, including a detailed description of the present weather [17]. In this paper, we focus on the task of nowcasting precipitation amounts in the near future for the purpose of generating alerts on extreme weather events and preventing damage to infrastructure due to flooding. Previous work on precipitation nowcasting has shown that even at small lead times of less than an hour, accurate precipitation prediction can be challenging. In fact, precipitation is one of the most difficult weather variables to accurately forecast [26].

Traditional nowcasting techniques rely on radar reflectivity data and optical flow-based extrapolation methods to estimate the future position of precipitation fields. While computationally efficient, these methods struggle with predicting the initiation, growth, or dissipation of rain cells, particularly during convective storms. In recent years, deep learning has emerged as a promising approach to address these limitations by directly learning the complex spatiotemporal dynamics of precipitation from sequences of radar observations.

This paper investigates the potential of incorporating additional vertical atmospheric structure information — specifically, echo top height (ETH) — into radar-based deep learning nowcasting models. ETH is a variable derived from volumetric radar scans representing the top altitude of detected precipitation that may offer insight into the vertical development of rain cells, potentially enabling earlier detection of intensifying convection. We begin by exploring the relationship between ETH and radar reflectivity (dBZ) and establish that ETH contains complementary, though not redundant, information.

To assess the utility of ETH, we implemented several variants of U-Net architectures that incorporate the ETH variable in different ways. Among the tested models, we identified a 3D U-Net architecture as being the most suitable model for jointly learning from radar reflectivity and ETH data. This particular architecture allows to represent radar reflectivity and ETH as separate data channels and provides a natural way to incorporate information about the vertical structure of precipitation.

The remainder of this paper is structured as follows. Section 2 reviews the relevant literature on deep learning using precipitation nowcasting and positions our work within the broader research context. Section 3 focuses on the exploration of echo top height as an additional model input. Section 4 describes our proposed methodology in detail, including the model architecture and training procedure. Section 5 presents the experimental setup, evaluation metrics, and results, followed by a discussion of key findings. In Sect. 6, we summarize our contributions and suggest directions for future work.

2 Precipitation Nowcasting – An Overview

The short-term prediction of precipitation intensity and location – called precipitation nowcasting – is paramount for many critical applications. It is crucial for taking effective measures aimed towards disaster prevention and mitigation and making decisions in agriculture, transportation and other weather dependent domains. It is increasingly vital when facing the climate crisis which increases the intensity and variability of precipitation [29].

The short lead times in nowcasting strongly limit the type of data and methods that can be leveraged. Mid-to-long-term forecasts are operationally calculated by simulating the future state of the atmosphere using numerical weather prediction (NWP) models. However, these models are computationally expensive to run and cannot be used to quickly assimilate new data due to their large spin-up times [15]. By the time the simulation outputs are available, we may have already overshot the short-term nowcast target. Consequently, nowcasting methods mostly rely on simpler, data-driven methods that approximate the physical processes in the atmosphere but are fast to run and can be updated very often (e.g., every 5 min).

For precipitation nowcasting, the most common data sources are ground radar reflectivity observations. Weather radar can be used to monitor precipitation with high spatial detail and a high update rates of just a few minutes [17], making them the best fit for the task.

Since radar data are commonly mapped to a Cartesian 2-dimensional grid, many methods from the field of computer vision are applicable. For example, autoregressive models can be used to estimate the spatial correlation structure of past and current precipitation fields, from which the next images can be predicted, similar to the next frame video prediction problem. Typically, this also involves a motion estimation and extrapolation step, during which precipitating cells are advected along the principal direction of motion. The motion field itself can be estimated from an optical flow or cross-correlation algorithm, such as in TITAN [5], COTREC [14], STEPS [3] and others. A major shortcoming of these extrapolation-based nowcasting methods is that the precipitation field is assumed to only change very slowly over time as it moves over different areas, which may not be true in very dynamic weather conditions such as convective rain. While modern nowcasting approaches try to tackle this in various ways, there are fundamental limits to what can be predicted based on radar data alone, and most dynamic processes such as the growth, intensification and decay of rain cells, or more generally, any change in rain cell size or motion, currently cannot be reliably predicted [20].

In the past years, deep-learning approaches have started to gain attention by attempting to mitigate these main limitations and learn to predict the complex space-time dynamics of precipitation fields. Deep learning models can learn complex patterns while still satisfying the requirement of producing an output quickly, offering the potential to overcome many of the limitations inherent in traditional methodologies. However, these approaches come with their own set of challenges.

2.1 Deep Learning for Precipitation Nowcasting

Within the realm of deep learning for precipitation nowcasting, the problem is often framed as a spatiotemporal sequence prediction task, where the goal is to forecast future radar precipitation maps based on a sequence of past observations. This perspective allows researchers to adapt and apply various deep learning architectures that have proven successful in video prediction and other sequence modeling domains.

The first notable deep learning model used for precipitation nowcasting was the ConvLSTM model presented in [24], which integrates convolutional layers into the recurrent LSTM architecture, enabling the model to effectively capture both spatial and temporal correlations within successive radar precipitation maps. A limitation of this approach is the complex motion of precipitation over time that results in location variance of precipitation patterns in the successive observations (the corresponding precipitation object moves to a different part of the successive image), which the location-invariant convolution filters are inadequate for. The Traj-GRU model from [25] builds upon the ConvLSTM and solves the limitation by allowing the model to learn location-variant recurrent connections dynamically. It employs a subnetwork that generates flow fields determining the sampling locations for the hidden states dynamically and allows TrajGRU to capture complex spatiotemporal correlations more effectively than ConvLSTM.

Despite the sequential character of the data, some researchers decided to abandon the recurrent neural network architecture for simpler convolutional-only models. A widely used architecture is U-Net – an encoderdecoder structure with skip connections. Its first notable use for precipitation nowcasting was RainNet in [1], treating nowcasting as an image-to-image translation problem. RainNet is trained to predict only a single next observation. However, longer forecasts can also be obtained by recursively applying the same model and using previous predictions as inputs. Previous research has shown that this can lead to cumulative smoothing effects, and a general loss of detail in predicted precipitation patterns over time and intensity degradation [1]. However, these limitations can be mitigated by dropping the iterative approach and predicting all the required lead times at once.

The blurriness issue is not limited to RainNet. It affects the vast majority of machine learning-based precipitation nowcasting models that generate a single deterministic prediction. The increasingly blurry output with increasing lead time is a major issue when trying to forecast heavy rain. The root cause is the usage of gridpoint-based error metrics as loss functions to minimize, which results in the so-called 'double penalty problem'. A forecast of a precipitation feature that is correct in terms of intensity, size, and timing, but incorrect in its location, causes a very large error [10]. This heavily interferes with the main goal of the nowcasting models – prediction of heavy precipitation events associated with significant societal impacts – as the highest intensity values in the data degrade over time.

The blurriness problem, inherent to deterministic models predicting the evolution of highly chaotic and uncertain precipitation fields, led to the introduction of adversarial learning approaches (GANs) such as in the DGMR model [21]. The latter aim to create visually plausible outputs without blurring over time by incorporating one or several discriminators. Another useful addition is to incorporate the underlying physical laws governing the behavior of rain through so-called "physics-informed machine learning" like in [30]. Such models are trained to predict dynamic extrapolation motion fields similar to the TrajGRU, while making sure that the predictions satisfy some basic physical or mathematical property such as the continuity equation. This ensures that the outputs are not just visually plausible, but also physically consistent with the laws of nature.

However, just like deterministic models can produce blurred predictions, GANs are prone to many issues, such as training instability and the generation of artifacts. Some models like CasCast [7] produce a deterministic prediction that then serves as an input to a generative diffusion model, thereby trying to combine the two approaches by capitalizing on their respective strengths while minimizing their downsides.

While deep learning methods have pushed the limits of what we thought was possible in operational rainfall nowcasting [6,30], many fundamental issues remain. Modeling the complex evolution of precipitation systems across different spatial and temporal scales and achieving accurate forecasts for extreme precipitation events remain key areas of ongoing research. To address these limitations, there is a growing interest in incorporating additional physical constraints and data sources into deep learning models, to help them better predict the nonlinear dynamics of rain cells.

2.2 The Potential of Additional Data Variables

To obtain a more accurate nowcast at higher lead times, additional attributes about the weather not captured by the radar reflectivity maps are helpful [11]. Key atmospheric variables such as air pressure, wind and temperature which heavily affect future weather evolution should be considered. Presumably cloud type, vertical profiles of temperature and pressure would help as well, but it's not clear how much.

Additionally, there are many factors that affect the utility of additional variables. First, to include data from other atmospheric sensors, we might need to perform computationally expensive data assimilation, which represents a sizable proportion of the cost of producing a forecast [2]. For the task of nowcasting, we need to carefully balance the benefits that the additional data sources can provide with the cost of their inclusion at operational runtime (gathering the various observations and their spatio-temporal mapping). An equally important aspect to be considered is the fact that surface data on their own are insufficient to characterize and predict the dynamics of clouds and precipitation. Unfortunately, the availability of vertical profiles and 3D atmospheric observations is limited, especially in real-time.

Another issue is the fact that the relationship between the primary and secondary variables in the model might change over time, depending on other, hidden factors or variables. Measurement errors and sampling uncertainty might also be an issue. Additionally, additional input channels actually increase the risk of overfitting and cause the predictions to become less reliable [8]. The prediction error might even be slightly lower but the fluctuations of the prediction errors over time and across events might be larger than before, which is undesirable.

Another crucial point deserves attention. Most machine learning approaches to nowcasting implicitly assume a fixed, stationary relationship between inputs and outputs. However, in the context of precipitation nowcasting - where chaotic, highly dynamic atmospheric processes are at play - this assumption often breaks down. Key information necessary to accurately predict future developments is often missing or unobservable. As a result, while a model might successfully learn the average mapping between inputs and outputs over a training dataset, this has limited practical value if it cannot adapt to individual, especially extreme, cases. These rare events often fall outside the learned average behavior, yet they are precisely the ones that need to be forecasted accurately.

2.3 Synthesis

Precipitation nowcasting predicts rain in the near future and is critical for timely decisions in weather-sensitive sectors. Traditional NWP models are too slow for short-term use, so fast methods working with radar observations are preferred. Deep learning models can be used for nowcasting, but they struggle with issues like blurriness, instability, and performance issues on rare extreme events. Incorporating extra atmospheric data may help but may also confuse the models.

3 Radar Reflectivity and Echo Top Height

Most existing radar-based nowcasting approaches rely solely on 2D radar reflectivity representations, such as constant altitude plan position indicators (CAPPI) or vertically integrated maxima (CMAX). While effective, these projections discard much of the vertical information captured by weather radars, which operate by scanning at multiple elevation angles and thus provide a full 3D view of the precipitation structure surrounding the radar station. Recent work [19] has demonstrated that incorporating this full 3D volumetric data can improve nowcasting accuracy (Fig. 1).

However, working with a full 3D volume can be very computationally expensive. Also, if we consider the time steps as a separate data dimension, we effectively arrive at 4D data volumes that the model needs to process and many popular machine learning frameworks do not even support 4D convolution operations natively. As a result, there is a need for compact yet informative representations of vertical storm structure that can be more easily integrated into existing 2D-based models. This is where echo top height comes in.

Echo top height (ETH) refers to the maximum altitude at which a weather radar detects a reflectivity value above a certain threshold. This measurement serves as an indicator of the storm's vertical extent and intensity, and can be used to provide insight into a storm's development and severity. Various reflectivity thresholds such as 7 or 18 dBZ are used depending on the specific application and radar system. See Fig. 2 for an example observation.

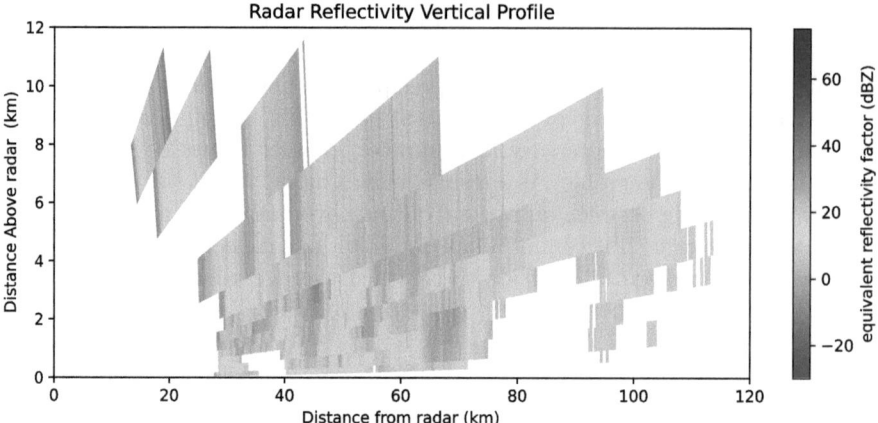

Fig. 1. This Example shows a single vertical slice of the full 3-dimensional radar observation. The radar station is located in the bottom left corner. Each ray represents data captured by a different radar sweep.

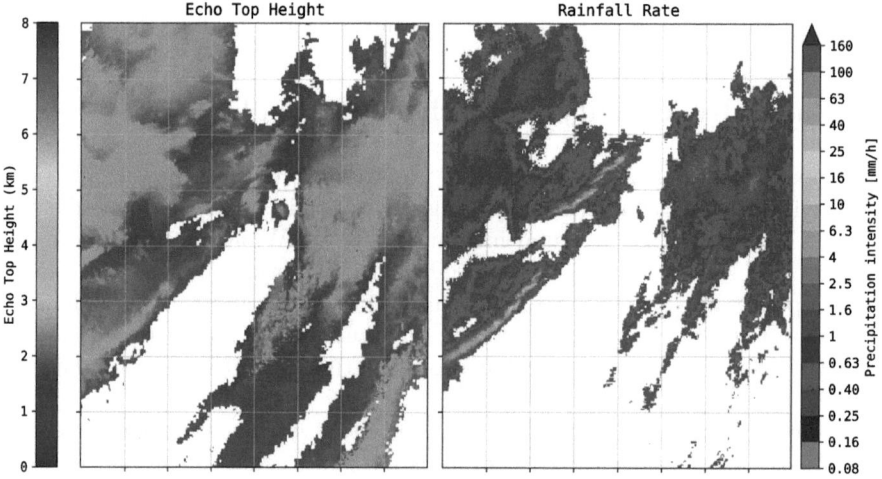

Fig. 2. Side-by-side comparison of a single observation from the dataset (February 20th, 2022), showing the spatial distribution of echo top height (left) and corresponding rainfall rate (right). This illustrates the relationship between storm structure aloft and precipitation intensity at 1500 m above ground. In this case, a reflectivity threshold of 7 dBZ was used to calculate echo top height

3.1 Relationship of Echo Top Heights and Reflectivity

To illustrate the relationship of echo top heights and reflectivity, we plot the co-occurrences of radar reflectivity and echo top height pairs of observations from three days of interest in Fig. 3 (we use data from De Bilt and Den Helder stations in Netherlands, see Sect. 4.1 for details about the data used). A strong relationship is visible in all of the plots – the bottom right part is empty, meaning there are almost no observations with high rainfall rate and low echo top height. This implies that high values of echo top height could help us identify the extreme precipitation events – as one variable increases, the other also increases proportionally. However, the correlations do not seem to be so strong that adding the echo top height variable would be meaningless.

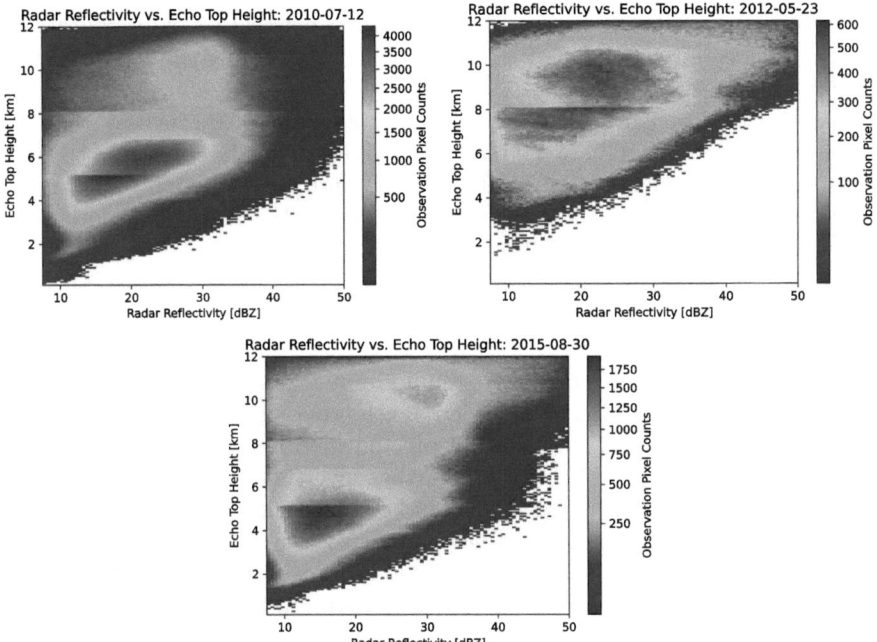

Fig. 3. The distribution of radar reflectivity with their corresponding echo top height observations for three different days with strong precipitation. The plots suggest a roughly proportional relationship between the two variables. The echo top data also clearly contains artifacts corresponding to the maximum observable echo top height of various radar sweeps, visible as density cutoffs at various heights.

3.2 Related Uses of Echo Top Heights

Echo top height (ETH) has served as a valuable resource in many applications. Higher ETH values generally indicate stronger updrafts and more intense convection, often associated with severe weather phenomena like heavy precipitation, hail, and lightning. In aviation, ETH is a critical parameter for assessing potential hazards to aircraft [9]. Furthermore, ETH has been investigated for its utility in predicting lightning initiation, with studies suggesting that ETH reaching at least 7 km are a necessary condition for cloud-to-ground lightning strikes [28], showing a connection to lightning initiation during storms.

Notably, ETH has also been explored as an indicator of rainfall rate incorporated into quantitative precipitation estimation – the task of mapping radar observations to the precipitation on ground. In [31], the use of ETH in conjunction with radar reflectivity (Z) for quantitative precipitation estimation using a GRU neural network was explored. This research demonstrated that the inclusion of ETH as an input feature significantly improved the accuracy of rainfall estimation compared to using only radar reflectivity. Although this study concentrated on precipitation estimation rather than nowcasting, it provides compelling evidence for the value of ETH in a machine learning context related to precipitation analysis.

Likewise, it was shown that echo top height is one of the factors that can influence the relationship between radar reflectivity (Z) and rainfall rate (R) [27]. This suggests that incorporating ETH directly into a nowcasting model might help to implicitly account for the variability in the Z-R relationship, potentially leading to more accurate rainfall predictions.

3.3 Synthesis

Based on the reviewed material, there is a noticeable gap in the current scientific literature regarding the direct utilization of echo top heights (ETH) as an input channel within deep learning models for precipitation nowcasting. The potential benefits of incorporating ETH warrant further investigation.

It is plausible that including ETH could enhance the model's ability to represent the vertical structure and intensity of convective systems more effectively than using reflectivity data alone. Changes in ETH over time might also provide valuable cues for predicting the initiation, growth, and decay of precipitation, particularly in convective events. Furthermore, its correlation with intense convection suggests it could improve the nowcasting of heavy precipitation. However, it is possible that the information provided by ETH is not beneficial and there is a significant degree of redundancy when used alongside the radar reflectivity data.

In conclusion, the analysis of the provided scientific literature indicates that leveraging echo top heights in deep learning models for precipitation nowcasting represents a promising, yet relatively underexplored, research avenue.

4 Methodology

To evaluate the potential benefits of incorporating echo top height (ETH) in deep learning-based precipitation nowcasting, a suitable neural network architecture must be selected. While recent work in the field is increasingly focused on generative approaches, these models are computationally expensive to train and evaluate. Moreover, achieving state-of-the-art performance is not the primary aim of this study; rather, our focus is on proof-of-concept experimentation.

Given these considerations, we opt for a deterministic model that produces a single nowcast rather than a distribution of possible outcomes. Although deterministic models are affected by the so-called double penalty problem – manifesting as spatial blurring and intensity degradation over time – we consider this a potential advantage in our context. If the inclusion of ETH effectively reduces uncertainty in the predicted evolution of precipitation, any improvement should be visible as a reduction in this blurring effect.

We specifically selected the U-Net architecture as the neural network architecture experiment for its simplicity and wide use in nowcasting. See Sect. 4.2 for more details on the U-Net architecture used and the training setup. We use mean squared error as a loss function to minimize.

To ensure our experimental results are not due to random chance, we train multiple models using different random seeds and varying train-validation splits. This helps assess the consistency and robustness of the models across different initializations and data subsets. In addition to visual, qualitative evaluation on selected nowcasting events, we use a combination of general pixel-wise metrics, threshold-based metrics over binarized maps and also include the Fractions Skill Score (FSS), a very popular metric used in forecasting which evaluates spatial accuracy at different scales and accounts for slight spatial misalignments in precipitation fields [22].

4.1 Dataset

The data used for training and validating the model consists of radar observations for the Netherlands provided by the KNMI [23]. The reflectivity data can be downloaded from the KNMI data platform at this link [13] and the corresponding echo top heights here [12]. The datasets were generated by combining the data from two C-band radars in De Bilt (now in Herwijnen) and Den Helder at various elevation angles. For the 1500 m reflectivity product, a few low-elevation angle scans are interpolated and combined using a distance-weighted average. For echo top height products, more elevation angles are used to find the highest altitude where reflectivity exceeds 7 dBZ, and the maximum value from both radars is taken. In both cases, the file timestamp reflects the start of the lowest scan and the data are updated at regular 5 min intervals.

The whole dataset spans a 15-year period between 2008 and 2022, comprising over 1.5 million observations at a 5-minute temporal resolution, excluding occasional missing data. The majority of these observations represent clear-sky conditions, which, if included, would introduce a significant bias toward low-precipitation events. To address this imbalance, we computed a precipitation event weight for each observation by summing the squared radar reflectivity values across all pixels. Then, we selected the top 1000 observations from each year and used them as the starting points of dataset sequences. Each training sequence consists of 22 consecutive observations, which results in 15000 overlapping sequences consisting of 24118 total selected observations. In total, only around 1.5% of the total available data were included in the dataset for training and evaluating the model.

However, during preprocessing, we identified a significant distributional shift in the echo top height data beginning in the second half of 2016, coinciding with the installation of new polarimetric radars by KNMI. Therefore, to ensure data consistency, we restricted model training to data collected from October 2016 onward.

In terms of spatial resolution, each pixel in the dataset corresponds to a 1×1 km area. The full radar image spans 765×700 km. However, a significant portion of the image lies beyond the effective range of the radars. This limitation becomes particularly evident in the echo top height data. Since the radar performs sweeps at multiple elevation angles, the maximum detectable altitude varies across the image. Near the edges of the coverage area, which is only reached by the lowest-angle sweeps, the radar cannot detect echo tops above approximately 2 km. In contrast, the echo top heights at the center of the image can reach up to 16 km. To ensure consistent vertical coverage across the input data, we restricted the training dataset to a central subset of the image (336×272 km), where echo tops of at least 7 km are detectable across the entire region. The final spatial extent used for training is shown in Fig. 4. While this still results in considerable artifacts present in the echo top data when objects travel between sweeps, a smaller extent would be undesirable as it would drastically reduce the spatial coverage and amount of training data.

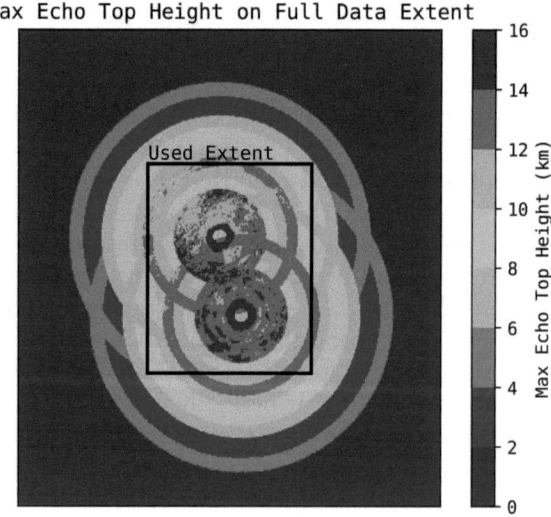

Fig. 4. Maximum observed echo top height per pixel across all training and test set samples (October 2016 to December 2022). The highlighted area indicates the spatial extent selected for model training, ensuring vertical coverage of at least 7 km.

For the model training and evaluation, we convert the radar reflectivity to rainfall rate using the Marshall-Palmer formula [16]. To mitigate speckle-like noise in the radar reflectivity data, we apply a morphological clutter removal procedure to each frame of the radar composite, following the approach used in [18]. First, the reflectivity field is thresholded to create a binary mask, where pixels corresponding to rain rates above 0.1 mm/h are set to 1, and all others to 0. This binary mask is then processed using a morphological opening operation, consisting of three iterations of erosion followed by three iterations of dilation. This procedure effectively removes small, isolated reflectivity regions that are likely the result of measurement noise or non-precipitating clutter.

As a test set for evaluation, we use the sequences from the last year of the dataset – 2022. Besides using these for quantitative evaluation, we also picked three precipitation events of interest for qualitative evaluation purposes in Sect. 5.4.

4.2 Architecture Exploration and Preliminary Experiments

We conducted a series of preliminary experiments to explore how echo top height (ETH) data could be integrated into deep learning-based precipitation nowcasting. As a foundation, we implemented a variant of the RainNet U-Net architecture proposed by Ayzel et al. [1], which was extended in several directions to assess the effect of ETH and different modeling strategies.

To incorporate ETH, we first experimented with a straightforward extension of the original 2D convolutional model by interleaving ETH and rainfall observations as additional input channels. However, this conflated temporal, and variable dimensions. To address this, we restructured the input to treat ETH and rainfall as separate channels across time, leading to a natural use of 3D convolutions in the U-Net (where time is the third dimension) that could now model spatial and temporal dependencies jointly.

Early results suggested that both the inclusion of ETH and the shift to 3D convolutions improved short-term prediction accuracy. However, we observed increasing systematic bias at longer lead times in models since they used recursive input strategy where past predictions feed into future forecasts.

To mitigate this, we tried constraining ETH outputs during training by incorporating a secondary loss term. While this reduced bias accumulation, it also degraded overall forecast performance. As an alternative, we modified the architecture to predict all future frames in a single forward pass, removing the recursive loop. We chose to predict 18 future frames at once, resulting in nowcasts from 5 to 90 min into the future. This approach eliminated the long-term bias issue and maintained comparable accuracy to the iterative approach.

In short, initial experiments showed that incorporating ETH as an additional input and adopting 3D convolutions (treating time as a separate dimension) improved predictive accuracy. However, these models exhibited increasing systematic bias at longer lead times due to recursive input dependencies. Attempts to mitigate this bias by constraining ETH predictions during training reduced the bias but degraded overall accuracy. An alternative approach—removing the recursive strategy and instead directly predicting all future frames—eliminated the bias and maintained competitive accuracy.

4.3 Synthesis

The study opts for a deterministic, U-Net-based architecture for simplicity and interpretability, using mean squared error as the loss function. Despite its susceptibility to blurring over time, the model's performance degradation is considered useful for identifying ETH's effect in reducing prediction uncertainty. The dataset comprises radar reflectivity and ETH data from the Dutch KNMI, spanning 20082022. However, due to changes in radar hardware in 2016, to ensure data consistency, only post-October 2016 data is used. The final model used in this study is a 3D U-Net variant that processes spatiotemporal data using 3D convolutions and predicts all 18 future precipitation frames in a single forward pass, avoiding recursive prediction and reducing long-term bias accumulation.

5 Evaluation

To ensure the robustness and reliability of our experimental results — and to rule out the possibility that they are due to random chance or favorable initial conditions - we trained eight instances of our models, both with and without echo top height inputs while using different random seeds as well as varying train-validation data splits. Each model used roughly one eighth of the training sequences as validation set for early stopping and learning rate scheduling. All the models were evaluated on the same test set, that is, sequences from the last year of the dataset in 2022.

Results are assessed both quantitatively and qualitatively. On the qualitative side, we visually inspected the forecasts on selected events to assess how well the different models capture spatial and temporal patterns in precipitation evolution, particularly in cases with heavy rain and complex rainfall dynamics. On the quantitative side, we use a combination of general-purpose pixel-wise error metrics and domain-specific verification scores. The pixel-wise metrics include Mean Absolute Error (MAE), which measures the average magnitude of errors regardless of direction; Mean Squared Error (MSE), which penalizes larger errors more severely and is sensitive to outliers; and Mean Error (ME), also known as bias, which indicates whether the model systematically over- or underestimates precipitation. To gain additional insight about performance at various precipitation levels, we further include precision, recall, and the equitable threat score (ETS), which jointly assess detection skill, false alarms, and overall forecast accuracy for selected binary precipitation thresholds. In addition to these, we also considered the Fractions Skill Score (FSS), a widely-used metric in the nowcasting community that evaluates spatial consistency and skill at different spatial scales. The FSS is particularly useful in precipitation forecasting as it accounts for slight spatial misalignments between predicted and observed fields—something traditional pixel-wise metrics tend to penalize heavily, even if the forecast is meteorologically reasonable.

5.1 Quantitative Evaluation: MAE, MSE, ME

We begin by evaluating model performance using standard point-wise metrics: mean squared error (MSE), mean absolute error (MAE), and mean error (ME), also referred to as bias. To assess the variability across models, we computed the mean and standard deviation of each metric at all forecast lead times for two groups of models: eight trained with echo top height (ETH) input and eight without. This setup allows for a comparison of predictive accuracy and consistency between the two approaches. The aggregated results are presented in Fig. 5.

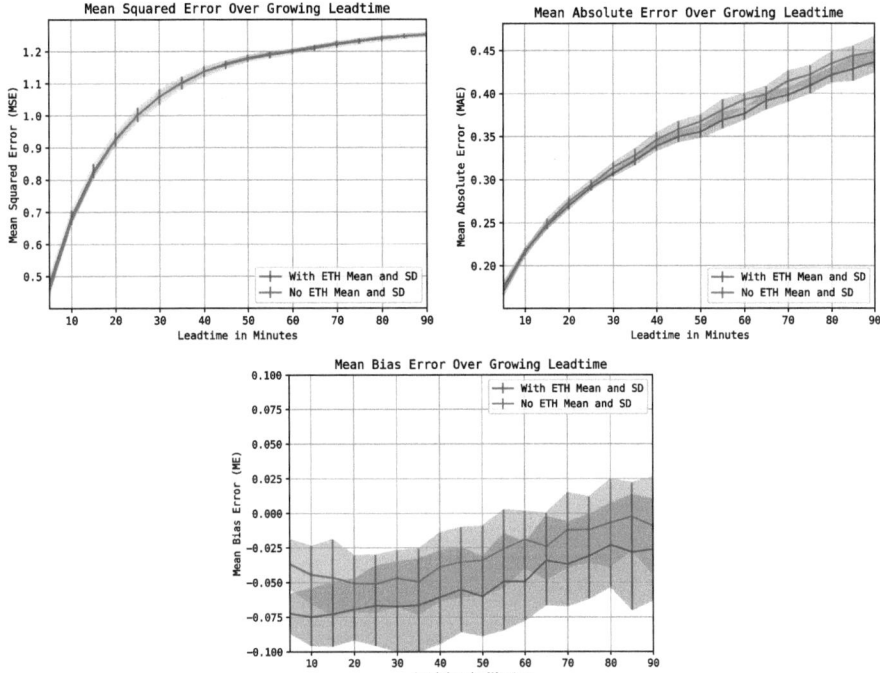

Fig. 5. The means and standard deviations of models with and without echo top height input for three metrics – mean squared error (MSE), mean absolute error (MAE), and mean error (ME) or bias. Each group consists of 8 separate models trained with different test-validation splits and different random seeds for both the models with (green) and without (red) echo top heights. (Color figure online)

The MSE results indicate nearly identical average errors between the two groups, though the ETH-based models display greater variability. The MAE curves show that models with ETH consistently perform better at longer lead times, suggesting improved stability in the nowcasts. However, the ME plots reveal a more pronounced negative bias in the ETH-enhanced models, suggesting a systematic tendency to underestimate precipitation compared to their non-ETH counterparts.

The greater variance in MSE for the models with ETH suggests increased sensitivity. The outputs of the models that rely on ETH features vary more strongly across different training subsets, suggesting that performance is likely very case specific. Despite this, the improved MAE at longer lead times suggests that ETH provides useful contextual information, even if the absolute precipitation levels are occasionally underestimated.

To further investigate the conditions under which the models perform better or worse, we analyzed the test set samples as a function of observed precipitation intensity. Specifically, each sample was ranked according to its maximum and spatially averaged precipitation rates at 30 min lead time, providing a two-dimensional representation of event severity. We then computed the mean difference in evaluation metrics between the two model groups with and without echo top height (ETH), and used this difference to color each sample accordingly. This visualization, shown in Fig. 6, illustrates how the relative performance of ETH-enhanced models varies across different precipitation events.

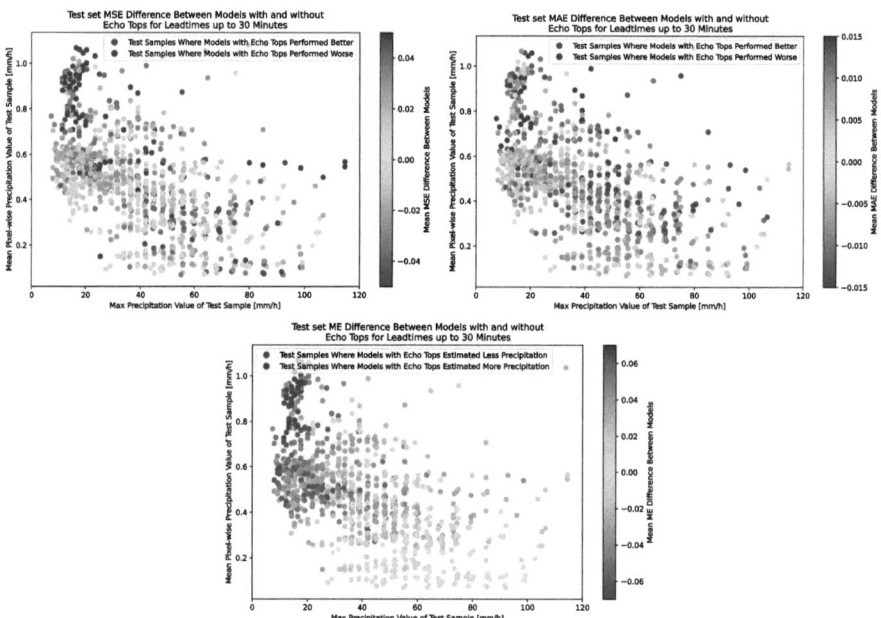

Fig. 6. Mean difference of models with and without echo top height input for three metrics – mean squared error (MSE), mean absolute error (MAE), and mean error (ME) or bias – for each test sample based on their maximum and spatially averaged precipitation rate.

The MSE and MAE plots reveal a distinct pattern: models with ETH input tend to perform worse on events with broad but moderate rainfall (i.e., higher average precipitation and lower peak intensities), yet perform better on events characterized by localized, high-intensity precipitation (i.e., higher maximum values and lower spatial averages). This suggests that ETH information helps the model focus on intense convective features but also introduces noise and uncertainty in more uniform rainfall conditions. The ME plot, on the other hand, shows no consistent relationship with either precipitation metric. The observed reduction in absolute bias for lower-rainfall events is expected, as such cases inherently limit the range of possible prediction errors.

5.2 Quantitative Evaluation: Precision, Recall, ETS

To complement the pixel-wise error metrics and gain a more event-focused perspective on model performance, we evaluate the predictions using threshold-based categorical metrics: precision, recall, and equitable threat score (ETS). These metrics are particularly useful in assessing the ability of the models to correctly identify precipitation occurrences above specific intensity levels, which is critical for operational forecasting and decision-making. The evaluation is conducted using four precipitation rate thresholds – 0.1, 1, 2.5, and 5 mm/h – chosen to span a range from light drizzle to moderate rainfall. At each precipitation level, predicted and observed precipitation fields are thresholded into binary maps of 0 and 1, and the metrics are computed accordingly. Precision reflects the fraction of predicted precipitation events that were actually observed, recall measures the proportion of observed events that were correctly predicted, and ETS provides a balanced assessment that accounts for random hits. This thresholded analysis allows for a more nuanced understanding of how the inclusion of echo top height input affects the models' capacity to predict precipitation with varying intensity. See Figs. 7, 8, and 9 for the plots.

Models that utilize ETH input consistently achieve higher precision, particularly at lower thresholds, indicating a tendency of the models toward predicting less rainfall with fewer false alarms in light precipitation scenarios. Conversely, models without ETH input exhibit higher recall across all thresholds, reflecting a greater ability to detect precipitation events, though at the cost of an increased rate of false positives. The ETS, which balances precision and recall, favors ETH-based models at the lowest threshold (0.1 mm/h). However, at higher thresholds, models without ETH input demonstrate superior performance. These findings suggest that ETH information enhances the detection of weak precipitation signals but is less beneficial – or even detrimental – when trying to forecast higher rainfall intensities.

5.3 Quantitative Evaluation: Fractions Skill Score

To evaluate the spatial accuracy of predicted precipitation fields, we compute the Fractions Skill Score (FSS), a widely used metric in quantitative precipitation forecasting that accounts for both the intensity and spatial displacement of forecasted rainfall. Unlike pointwise metrics, FSS quantifies how well the predicted and observed precipitation fields align over neighborhood regions, making it suitable for assessing convective-scale forecasts where slight spatial shifts can

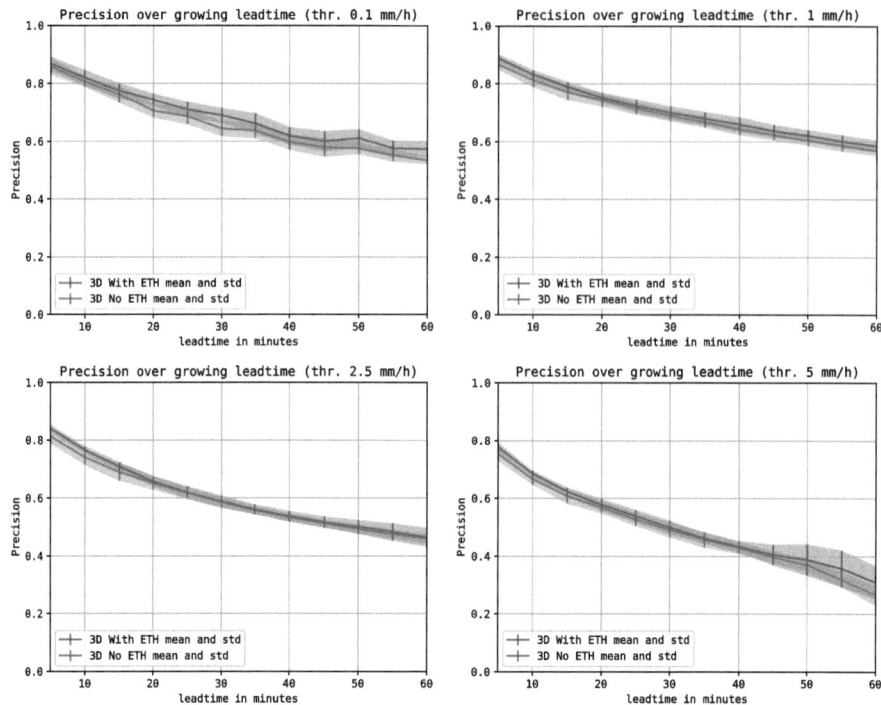

Fig. 7. Means and standard deviations of models with and without echo top height (ETH) for precision, evaluated at four rainfall intensity thresholds: 0.1, 1, 2.5, and 5 mm/h. Each group consists of 8 separate models trained with different test-validation splits and random seeds. Green indicates models with ETH; red indicates models without ETH. (Color figure online)

lead to large pointwise errors. The FSS is evaluated across five precipitation rate thresholds (0.1, 1, 2.5, 5, and 10 mm/h) to capture performance across a range of rainfall intensities. Additionally, to examine the effect of spatial scale, the scores are computed using three spatial tolerance radii corresponding to 1, 4, and 16 km. This multiscale, thresholded approach provides a nuanced understanding of the models' ability to capture both occurrence of precipitation events across different intensities and lead times. See Fig. 10 for the score matrices.

The results show that models incorporating echo top height (ETH) inputs achieve higher FSS values at lower precipitation thresholds, particularly for shorter lead times. This improvement is most evident at the 0.1 mm/h threshold, where ETH-based models consistently outperform those without ETH across all spatial scales. However, as the precipitation threshold increases, the differences in FSS between the two model configurations quickly diminish. At moderate to high thresholds (1 mm/h and above), the scores are similar, suggesting that the inclusion of ETH does not improve the spatial predictive skill for heavier rainfall events. From this, we conclude that the additional information provided by ETH

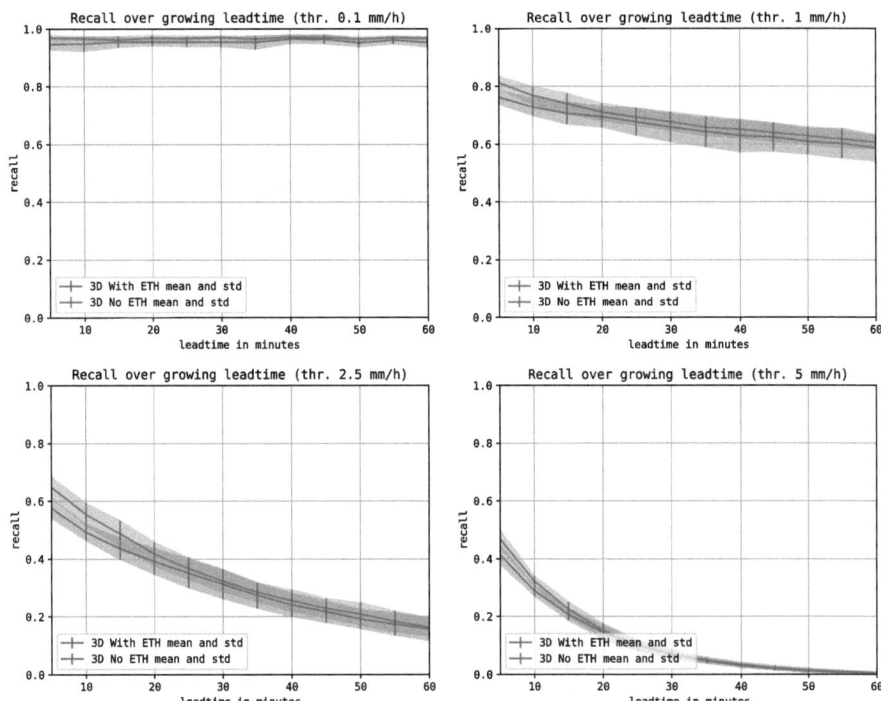

Fig. 8. Means and standard deviations of models with and without echo top height (ETH) for recall, evaluated at four rainfall intensity thresholds: 0.1, 1, 2.5, and 5 mm/h. Each group consists of 8 separate models trained with different test-validation splits and random seeds. Green indicates models with ETH; red indicates models without ETH. (Color figure online)

mostly contributes toward improving the spatial coherence of light precipitation forecasts, while its benefit is limited for more intense rainfall.

5.4 Qualitative Evaluation: Extreme Event Case Studies

Finally, we present qualitative examples of model outputs to complement the quantitative evaluation. As previously discussed, numerically assessing nowcasting performance is inherently challenging, as the optimal forecast often depends on specific application needs and may be subjectively interpreted. To gain further insight into model behavior, we selected three test samples for more in-depth visual analysis and comparison of the forecasts generated by models with and without echo top height input.

The selected events include:

- **February 20, 2022:** Following closely after Storm Eunice, Storm Franklin brought strong winds and heavy rainfall, exacerbating damage from previous storms and causing localized flooding.

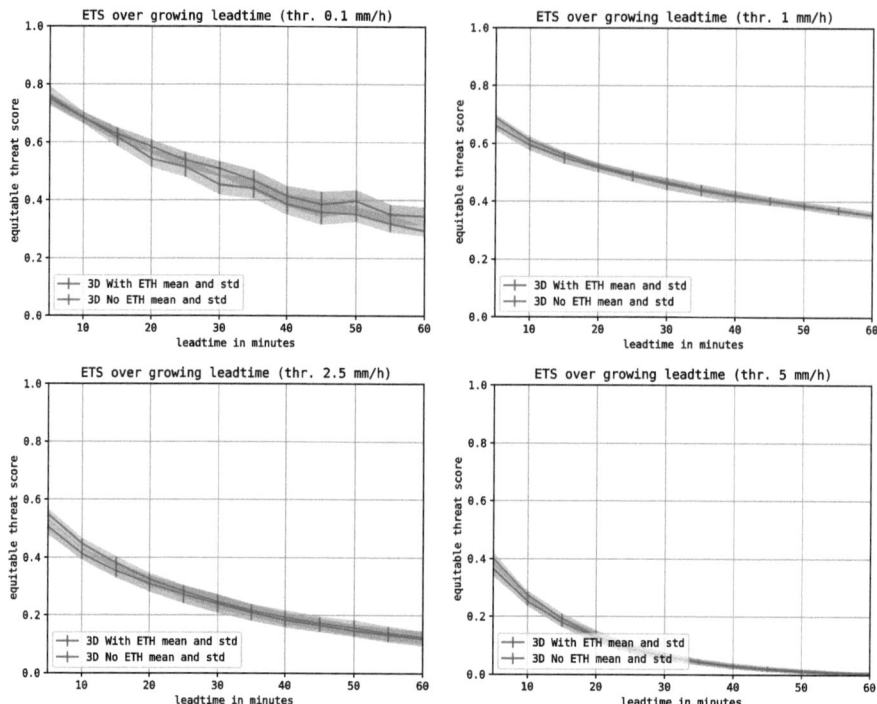

Fig. 9. Means and standard deviations of models with and without echo top height (ETH) for equitable threat score (ETS), evaluated at four rainfall intensity thresholds: 0.1, 1, 2.5, and 5 mm/h. Each group consists of 8 separate models trained with different test-validation splits and random seeds. Green indicates models with ETH; red indicates models without ETH. (Color figure online)

- **May 19, 2022:** On May 19, severe storms impacted the Netherlands, particularly affecting fruit farms in the Batavia region, leading to significant financial losses for fruit growers. Additionally, the southern province of Noord-Brabant and the city of Utrecht experienced flooding in homes and streets.
- **August 17, 2022:** On August 17, severe thunderstorms brought heavy rainfall to northeastern parts of the Netherlands, flooding streets and basements.

For the case studies above, we selected a single model from each group (i.e., with a without ETH) based on their performance over the entire test set. Specifically, we chose the model that achieved the highest fractions skill score (FSS) at a threshold of 2.5 mm/h, with a spatial tolerance scale of 16 km and a lead time of 30 min. As plotting each of the 16 trained model would be needlessly exhaustive, these were chosen to show the performance of "best" models of each group plotted side by side. The nowcasts generated by these selected models are visualized in Figs. 11, 12, and 13.

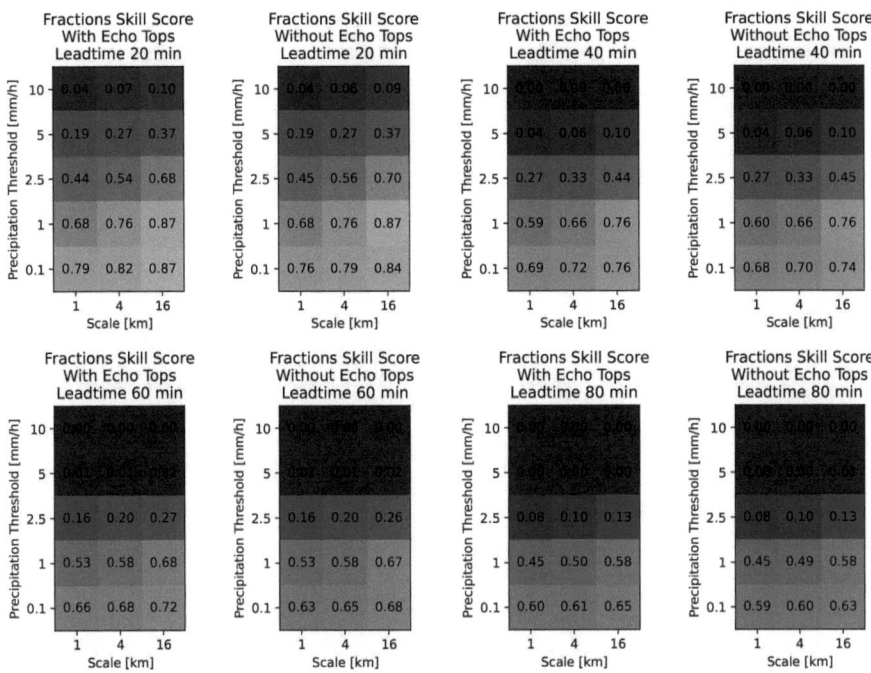

Fig. 10. The matrices show the mean Fraction Skill Score (FSS) for models trained with and without echo top height (ETH) input. Evaluations are performed at five rainfall intensity thresholds (0.1, 1, 2.5, 5, and 10 mm/h) and three spatial tolerance scales (1, 4, and 16 km), across lead times of 20, 40, 60, and 80 min.

Figure 11 illustrates a case featuring a well-defined band of precipitation propagating eastward. Notably, although the rainfall intensity is high, the corresponding echo top heights (ETH) remain relatively low across the affected areas. When comparing the nowcasts, the model incorporating ETH input shows a modest improvement in preserving rainfall intensity over time. However, the model without ETH input appears to capture the overall spatial structure and propagation of the precipitation band better, maintaining the coherence of the main precipitation feature for longer.

Figure 12 presents a case in which a region of intense precipitation, located near the center of the domain, propagates toward the northeast. In this instance, the associated echo top heights (ETH) are relatively high, exceeding 10 km in some areas. The model incorporating ETH input produces a visibly more accurate nowcast, exhibiting reduced intensity degradation, less spatial blurring over time, and good spatial alignment with the observed precipitation pattern in comparison to the model without ETH. In this case, the model was able to leverage the additional vertical structure information contained in the ETH to enhance nowcast quality.

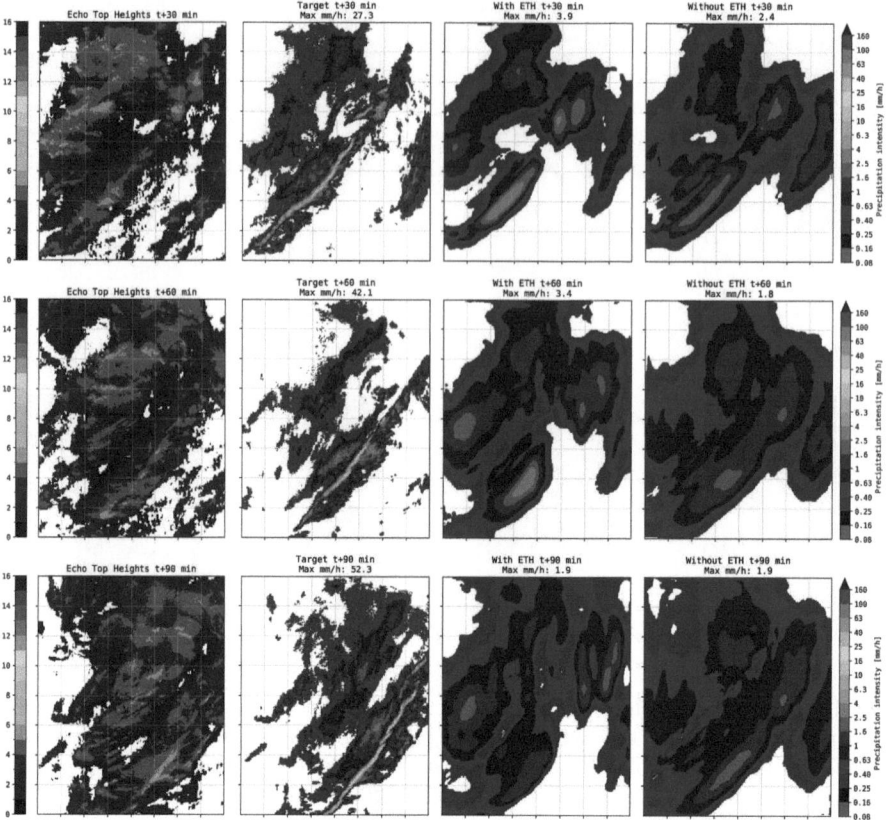

Fig. 11. Series of nowcasts generated by the models with and without ETH for a precipitation event on February 20th, 2022, at 20:00. For reference, the corresponding ETH observation and ground truth rainfall rate observation are also shown. Each panel presents the predicted or observed rainfall rates at successive lead times (T+30, +60, +90 min). To help with interpreting intensity trends, the maximum rainfall rate within each map is indicated numerically above the corresponding panel, providing a quantitative measure of nowcast rainfall intensity degradation over time.

Figure 13 depicts a case featuring a prominent precipitation system located near the northeastern edge of the domain. The corresponding echo top height (ETH) field exhibits artifacts – visible as circular bands – resulting from the separate radar sweeps having different ranges and maximum observation altitudes. Despite these limitations, the model with ETH input appears to effectively utilize the available ETH information, maintaining the intensity of the precipitation signal over a longer lead time horizon. The nowcasts also appear to be less spatially blurry. This interesting case suggests a certain degree of robustness in the model's ability to extract useful features from imperfect ETH observations.

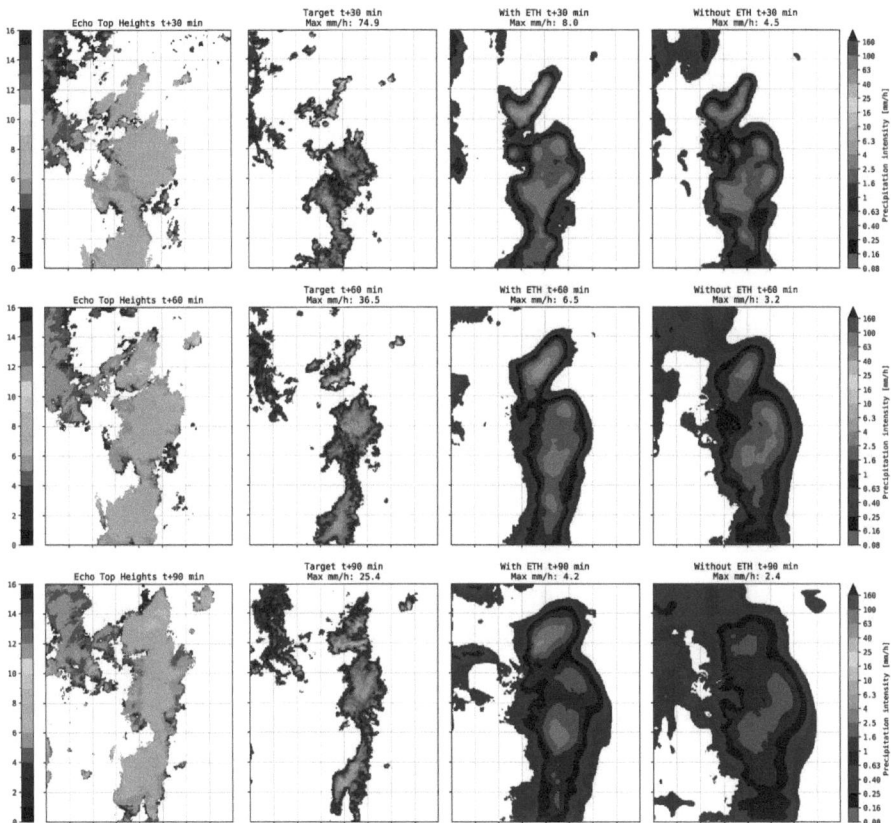

Fig. 12. Series of nowcasts generated by the models with and without ETH for a precipitation event on May 19th, 2022, at 03:00. For reference, the corresponding ETH observation and ground truth rainfall rate observation are also shown. Each panel presents the predicted or observed rainfall rates at successive lead times (T+30, 60, 90 min). To help with interpreting intensity trends, the maximum rainfall rate within each map is indicated numerically above the corresponding panel, providing a quantitative measure of nowcast rainfall intensity degradation over time.

While limited to only three cases, these qualitative comparisons offer valuable insight into how echo top height can influence nowcast performance under different precipitation scenarios. In the May case, where ETH values were high and coincided with intense convective activity, the model with ETH produced more accurate and spatially coherent forecasts. In contrast, the February case - with high rainfall but low ETH values - only showed modest gains and a less accurate depiction of the main precipitation structure, showing that ETH may not always be informative or helpful, and sometimes even detrimental to accurately predicting the space-time structures and dynamics.

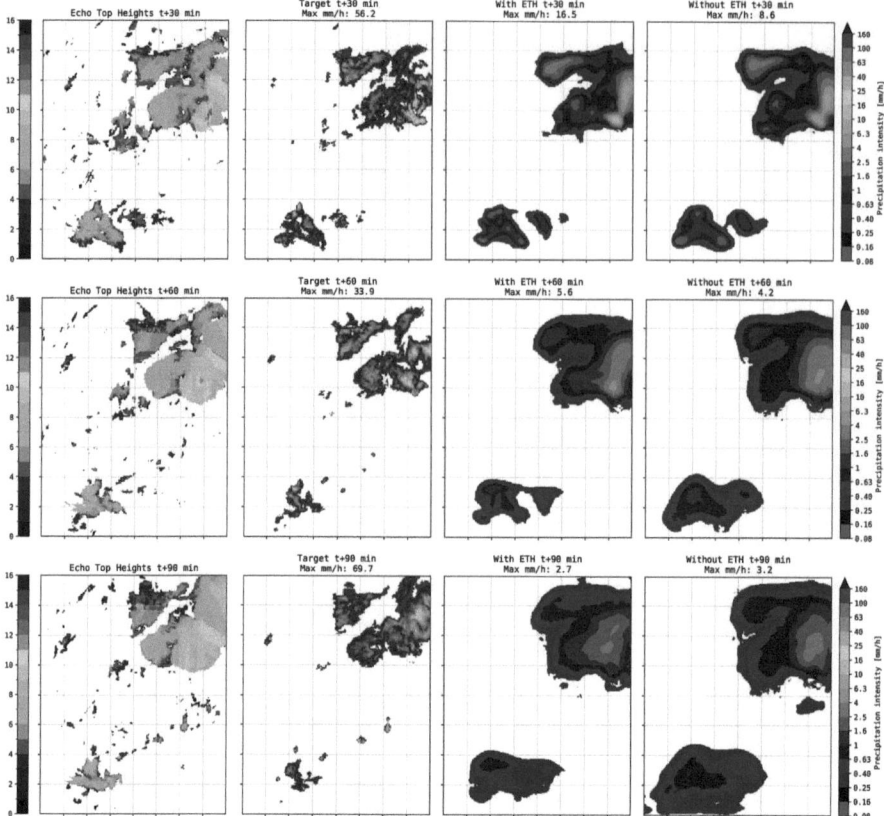

Fig. 13. Series of nowcasts generated by the models with and without ETH for a precipitation event on August 17th, 2022, at 15:00. For reference, the corresponding ETH observation and ground truth rainfall rate observation are also shown. Each panel presents the predicted or observed rainfall rates at successive lead times (T+30, 60, 90 min). To help with interpreting intensity trends, the maximum rainfall rate within each map is indicated numerically above the corresponding panel, providing a quantitative measure of nowcast rainfall intensity degradation over time.

The August case, despite containing obvious artifacts in ETH estimates, still showed some benefit in including ETH, retaining the precipitation intensity and spatial sharpness of the field over longer lead times. It also showed a certain degree of robustness of the model predictions to imperfect and noisy ETH data.

Taken together, these qualitative evaluations suggest that while echo top height (ETH) input does not universally and systematically improve nowcasts, it can provide meaningful benefits under the right meteorological conditions. The three case studies suggest that ETH might be particularly helpful in scenarios with high ETH observations. While not explored in this work, splitting the test set and analyzing model performance across different ETH ranges may offer further insight and are worth considering in future studies.

5.5 Synthesis

Eight model pairs with and without echo top height input were trained using varying random seeds and data splits. The results show that overall, incorporating ETH leads to slightly larger output variance and underestimation bias. While the use of ETH tends to improve predictive performance during light precipitation events, especially at short lead times, it generally worsens performance in cases of widespread moderate rainfall and at higher precipitation intensities. Qualitative visual evaluations of selected case studies further support these findings, revealing large case-specific differences in predicted spatial structure and intensity between model versions. These examples suggest that ETH can provide valuable contextual information under certain conditions. Future work should therefore focus on identifying the scenarios and better understanding the conditions under which the addition of ETH is most likely to improve the predictions.

6 Conclusion

In this study, we explored the potential of incorporating radar-derived echo top height variable (ETH) into deep learning-based precipitation nowcasting. Using a deterministic U-Net model with 3D convolutions, we showed that ETH can be effectively integrated as an additional input channel.

However, while the models were able to incorporate echo top height (ETH) data, the results do not provide convincing proof-of-concept evidence of its value for precipitation nowcasting. Improvements were mostly limited to very low rain thresholds, and there were inconsistent and unpredictable effects on reducing issues such as rainfall intensity degradation and spatial blurring. Additionally, the inclusion of ETH introduced an additional negative bias in the forecasts, which is undesirable for applications such as flood forecasting and early warnings.

These findings suggest that although ETH carries some physically relevant information, its usefulness for improving the prediction of near-surface precipitation processes in the Netherlands appears to be limited. More sophisticated model designs, better ETH preprocessing, introduction of generative or probabilistic approaches may be needed to fully exploit the potential benefit while mitigating the downsides. As such, our study offers a useful starting point and diagnostic benchmark for future efforts in improving deep learning-based nowcasting with 3D radar-derived features.

In a broader sense, this work highlights the challenges of leveraging additional information in radar-based nowcasting. We believe this work serves as both a valuable template and a cautionary tale for the critical evaluation of not only echo top heights, but also other auxiliary variables, such as dual-polarization radar products, prior to their operational integration.

Looking ahead, we plan to systematically analyze the impact of additional atmospheric variables on deep learning-based precipitation nowcasting. Building on the methods used in this study, the next step will involve looking for variables which provide consistent benefit across different weather conditions. Ultimately, this may lead us to explore multi-modal data fusion techniques that can effectively integrate diverse auxiliary inputs without compromising the nowcasting accuracy, hopefully providing a more stable and scalable approach to model design.

Acknowledgments. This research was partially supported by TAILOR, a project funded by EU Horizon 2020 research and innovation programme under GA No. 952215; by The Ministry of Education, Science, Research and Sport of the Slovak Republic under the Contract No. 0827/2021. Additionally, this research was partially supported by the Visegrad group for Vehicle to X (V4Grid), a project funded by Interreg Central Europe programme under CE0200803. The authors acknowledge the use of computational resources of the DelftBlue supercomputer, provided by Delft High Performance Computing Centre (https://www.tudelft.nl/dhpc [4]. Research results were obtained using the computational resources procured in the national project National competence centre for high performance computing (project code: 311070AKF2) funded by European Regional Development Fund, EU Structural Funds Informatization of society, Operational Program Integrated Infrastructure.

Online Visualizations. To better inspect the behavior and qualitative performance of the trained models, we refer the reader to a set of exported nowcast animations available at: https://pavlikp.github.io/ETOPS-results/.

Code Repository. All the training and evaluation code is available at https://github.com/kinit-sk/EchoTops.

Disclosure of Interests. The authors have no competing interests to declare that are relevant to the content of this article.

References

1. Ayzel, G., Scheffer, T., Heistermann, M.: RainNet v1. 0: a convolutional neural network for radar-based precipitation nowcasting. Geosci. Model Dev. **13**(6), 2631–2644 (2020)
2. Bauer, P., Thorpe, A., Brunet, G.: The quiet revolution of numerical weather prediction. Nature **525**(7567), 47–55 (2015)
3. Bowler, N.E., Pierce, C.E., Seed, A.W.: Steps: a probabilistic precipitation forecasting scheme which merges an extrapolation nowcast with downscaled NWP. Q. J. Royal Meteorol. Soc. J. Atmospheric Sci. Appl. Meteorol. Phys. T. **132**(620), 2127–2155 (2006)
4. Delft High Performance Computing Centre (DHPC): DelftBlue Supercomputer (Phase 2) (2024). https://www.tudelft.nl/dhpc/ark:/44463/DelftBluePhase2
5. Dixon, M., Wiener, G.: TITAN: thunderstorm identification, tracking, analysis, and nowcasting-a radar-based methodology. J. Atmos. Oceanic Tech. **10**(6), 785–797 (1993)

6. Espeholt, L., et al.: Deep learning for twelve hour precipitation forecasts. Nat. Commun. **13**(1), 1–10 (2022)
7. Gong, J., et al.: CasCast: skillful high-resolution precipitation nowcasting via cascaded modelling. arXiv preprint arXiv:2402.04290 (2024)
8. Guyon, I., Elisseeff, A.: An introduction to variable and feature selection. J. Mach. Learn. Res. **3**, 1157–1182 (2003)
9. Hwang, Y., Clark, A.J., Lakshmanan, V., Koch, S.E.: Improved nowcasts by blending extrapolation and model forecasts. Weather Forecast. **30**(5), 1201–1217 (2015)
10. Keil, C., Craig, G.C.: A displacement and amplitude score employing an optical flow technique. Weather Forecast. **24**(5), 1297–1308 (2009). https://doi.org/10.1175/2009WAF2222247.1
11. Klocek, S., et al.: MS-nowcasting: operational precipitation nowcasting with convolutional LSTMS at Microsoft weather. arXiv preprint arXiv:2111.09954 (2021)
12. Koninklijk Nederlands Meteorologisch Instituut (KNMI): Precipitation - radar 5 minute echo top height composites over The Netherlands (2024). https://dataplatform.knmi.nl/dataset/radar-echotopheight-5min-1-0. Accessed 22 Apr 2025
13. Koninklijk Nederlands Meteorologisch Instituut (KNMI): Precipitation - radar 5 minute reflectivity composites over The Netherlands - archive (2013). https://dataplatform.knmi.nl/dataset/radar-tar-refl-composites-1-0. Accessed 22 Apr 2025
14. Li, L., Schmid, W., Joss, J.: Nowcasting of motion and growth of precipitation with radar over a complex orography. J. Appl. Meteorol. Climatol. **34**(6), 1286–1300 (1995)
15. Liu, J., Bray, M., Han, D.: A study on WRF radar data assimilation for hydrological rainfall prediction. Hydrol. Earth Syst. Sci. **17**(8), 3095–3110 (2013)
16. Marshall, J.S., Palmer, W.M.K.: The distribution of raindrops with size. J. Atmospheric Sci. **5**(4), 165–166 (1948). https://doi.org/10.1175/1520-0469(1948)005⟨0165:TDORWS⟩2.0.CO;2
17. Organization, W.M.: Guidelines for nowcasting techniques. No. 978-92-63-11198-2, World Meteorological Organization, 7 bis, avenue de la Paix, P.O. Box 2300, CH-1211 Geneva 2, Switzerland (2017)
18. van Os, S.: Precipitation nowcasting using a generative adversarial network. Master's thesis, Delft University of Technology, Faculty of Civil Engineering & Geosciences (2024). http://resolver.tudelft.nl/uuid:35a11963-4524-442f-97a1-016694b3813f, mentor: M.A. Schleiss
19. Pavlík, P., Rozinajová, V., Ezzeddine, A.B.: Radar-based volumetric precipitation nowcasting: a 3D convolutional neural network with u-net architecture. In: CDCEO@ IJCAI, pp. 65–72 (2022)
20. Pulkkinen, S., et al.: PySteps: an open-source python library for probabilistic precipitation nowcasting (v1.0). Geosci. Model Dev. **12**(10), 4185–4219 (2019). https://doi.org/10.5194/gmd-12-4185-2019, https://gmd.copernicus.org/articles/12/4185/2019/
21. Ravuri, S., et al.: Skilful precipitation nowcasting using deep generative models of radar. Nature **597**(7878), 672–677 (2021)
22. Roberts, N.M., Lean, H.W.: Scale-selective verification of rainfall accumulations from high-resolution forecasts of convective events. Mon. Weather Rev. **136**(1), 78–97 (2008)
23. Royal Netherlands Meteorological Institute (KNMI): KNMI data platform. https://dataplatform.knmi.nl/ (2025). https://dataplatform.knmi.nl/. Accessed 22 Apr 2025

24. Shi, X., Chen, Z., Wang, H., Yeung, D.Y., Wong, W.K., Woo, W.: Convolutional LSTM network: a machine learning approach for precipitation nowcasting. In: Advances in Neural Information Processing Systems, vol. 28 (2015)
25. Shi, X., et al.: Deep learning for precipitation nowcasting: a benchmark and a new model. In: Advances in Neural Information Processing Systems, vol. 30 (2017)
26. Sun, J., et al.: Use of NWP for nowcasting convective precipitation: recent progress and challenges. Bull. Am. Meteor. Soc. **95**(3), 409–426 (2014)
27. Wu, W., Zou, H., Shan, J., Wu, S.: A dynamical Z-R relationship for precipitation estimation based on radar echo-top height classification. Adv. Meteorol. **2018**(1), 8202031 (2018)
28. Yang, Y.H., King, P.: Investigating the potential of using radar echo reflectivity to nowcast cloud-to-ground lightning initiation over Southern Ontario. Weather Forecast. **25**(4), 1235–1248 (2010)
29. Zhang, W., Zhou, T., Wu, P.: Anthropogenic amplification of precipitation variability over the past century. Science **385**(6707), 427–432 (2024). https://doi.org/10.1126/science.adp0212
30. Zhang, Y., et al.: Skilful nowcasting of extreme precipitation with NowcastNet. Nature **619**, 1–7 (2023)
31. Zou, H., Wu, S., Tian, M.: Radar quantitative precipitation estimation based on the gated recurrent unit neural network and echo-top data. Adv. Atmos. Sci. **40**(6), 1043–1057 (2023)

A Deep Dive into FAIRness Assessment: UReFM, a Formal Framework for Representing, Analyzing and Comparing Measures

Philippe Lamarre[1](✉)[iD], Jennie Andersen[2][iD], Alban Gaignard[3,4][iD], and Sylvie Cazalens[1][iD]

[1] INSA Lyon, CNRS, Ecole Centrale de Lyon, Université Claude Bernard Lyon 1, Université Lumière Lyon 2, LIRIS, UMR5205, 69621 Villeurbanne, France
{Philippe.Lamarre,Sylvie.Cazalens}@insa-lyon.fr
[2] Verkor, 2-4 rue Charles Bertier, 38000 Grenoble, France
[3] Nantes Université, CNRS, INSERM, l'institut du thorax, 44000 Nantes, France
alban.gaignard@univ-nantes.fr
[4] IFB-core, Institut Français de Bioinformatique (IFB), CNRS, INSERM, INRAE, CEA, 91057 Evry, France

Abstract. In recent years, the adoption of the FAIR principles has achieved notable success. This progress has led to the development of numerous assessment tools originating from diverse fields of application, thus addressing diverse object types, interpretations and implementations. Given the plethora of proposals available, it is crucial for users to precisely understand these measures, compare them effectively, make informed choices, and accurately interpret the obtained measurements. To meet these needs, we propose UReFM, a generic framework to formally represent and analyze measures with a tree like structure, FAIRness measures being a representative example. Besides the benefit of homogenization and of fine grained analysis, it allows for the formal definition of three characteristic quantities: coverage, granularity and impact. This latter reflects how much a given principle contributes to the final score. Our experiments show how these quantities (i) contribute to explain different scores obtained by digital artifacts using two different state-of-the-art assessment engines, (ii) enable a comparative study of different FAIRness measures, independently of any digital artifact.

1 Introduction

These recent years, the FAIR principles [49] have been increasingly adopted to assess the Findability, Accessibility, Interoperability and Reusability of their digital resources. Due to this widespread adoption, they have been specialized or even extended to meet the needs of very different scientific communities.

For instance, the RDA FAIR4RS working group derived these principles to specifically address research software [8], typically targeting their usability and reusability within other software. These principles have also been adapted in the context of AI [28], ontology development and semantic artifacts [12,40], data analysis workflows [46].

To support people in these assessment tasks, numerous tools have been developed, originating from diverse fields of application. They may address different types of objects, stem from different interpretations or implementations of the principles. Consequently, besides the varied terminology, one can notice that sometimes sub-principles are skipped, and how indicators are expressed or implemented changes from tool to tool. In addition, when scores are provided, they may not range in the same interval; the functions used to aggregate the scores, as well as the weights assigned to the various indicators, may also differ. In other words, the measurement methods, which we simply call *measures* all along this article, are diverse.

In some way, this diversity is understandable as the FAIR principles are not restrictive guidelines. However, with numerous and diverse available tools, a user may be faced to different questions such as: "Why does my resource get such a score with this tool?", "Why does it receive a higher score with tool A compared to tool B?", "What are the differences between tools A and B?", "Which one fits my needs better?", "Why is my digital resource considered better than this other one by tool X and worst by tool Y?". To answer such questions, some studies have already proposed comparisons of tools, based on metrics used, on characteristics of the tools themselves, on the measurements obtained with numerous datasets [10,23,31,37,42,43,51]. These studies are clearly useful, but it is still a challenge to interpret scores, understand and compare FAIRness measures, and make informed choices.

Indeed, we observe that the FAIR principles alone do not offer a sufficient framework to take into account the multiplicity of variations found from a measure to another. In addition, to our knowledge, there are no formally defined characteristic quantities to reflect the salient features of a given measure, that could help both its understanding and its comparison with other ones. More generally, the same problem arises each time principles are proposed to comply with some regulation such as the GDPR compliance checklist [22]. Several interpretations may result, and different tools assessing the compliance of digital objects may be implemented. Once more, a common framework to represent, understand and compare the underlying measures would be useful.

Hence our general goal is to define a formal and generic framework, enabling a uniform representation of measures organized according to a hierarchy of principles. This framework also aims at supporting the definition of descriptive quantities, in order to highlight salient traits of measures and facilitate their comparison. Another complementary goal is to confront the framework with real-life measures, both by modeling them and by conducting experiments showing how assessment scores can be better explained and how measures can be compared. The framework previously proposed [32] takes into account the FAIR principles only, limiting its generality. This paper extends this previous work and aims at defining a more general framework, with evolved intuitions and definitions. It in addition pays greater attention to the dependencies that may exist between the code of implementations, both from a theoretical and experimental point of view.

The remainder of this paper is organized as follows. In Sect. 3, following a discussion of related work, we introduce *UReFM* (*U*nified *Re*presentation *F*ramework for *M*easures), a generic formal model for representing measures. This model captures several key components: the principles structuring a measure, the indicators defining how to assess specific aspects, the programs implementing the indicators, possible dependencies between implementations, the resulting scores assigned to digital objects, and the methods used to aggregate these scores at the principle level. Section 4 illustrates the expressiveness of UReFM by using it to uniformly represent three representative FAIRness assessment tools: ARDC's tool [6], F-UJI [16], and FAIR-Checker [20]. This section also introduces a toy measure that serves as a running example throughout the paper. Section 5 defines key descriptive quantities—coverage, granularity, and impact—that support the understanding, analysis, and comparison of measures. To enable accurate comparisons across measures, Sect. 6 defines a set of operators that allows to align a given measure to a reference structure. We demonstrate this by aligning existing FAIRness measures to the FAIR Principles [49]. Section 7 presents an experimental study where several FAIRness measures are applied to a collection of digital objects. The used measures are compared using the introduced characteristic quantities, offering deeper insights into potential divergences between FAIR assessments. Finally, concluding remarks are provided in Sect. 8.

2 Related Work

The FAIR principles were published in 2016 [49] as general guidelines for the publication of digital resources to make them Findable, Accessible, Interoperable, and Reusable. Since then, numerous methods and tools have been developed to assess FAIRness, each with its own interpretation of the principles. This variety of interpretations and of types of assessment methods (automated tools, checklists, self-assessment questionnaires, etc.) makes them difficult to compare. To address this problem, the FAIR data maturity model [7] proposes a set of indicators that express measurable aspects of the FAIR principles and on which future evaluation tools can be based. Although this initiative proposes consensual definitions adopted by large multi-disciplinary communities, it is still challenging to compare FAIRness measures in their whole, from implementations to evaluation results.

Several comparisons have already been conducted. Slamkov *et al.* [42] compare five questionnaires and checklists: ARDC's tool [6], CSIRO's tool [13], SATIFYD [14], EUDAT checklist [29], and the SHARC grid [15], according to their main characteristics (type, documentation, dependency to a specific repository, automated score computation) and the results obtained on seven datasets.

Another comparison [43] focuses on three automated tools: F-UJI [16], FAIR Evaluation Service[1] [50] and FAIR-Checker [20]. They are all compared based on distinguishing aspects (documentation, availability of the code, format and log of the results...). Then the author focus on F-UJI and the FAIR Evaluation Service to compare their metrics/indicators in detail, first by focusing on their expression in natural language and then on the experimental results obtained on three datasets. Since then, both F-UJI and FAIR-Checker have changed. Wilkinson et al. [51] highlight that some of the differences between the results of F-UJI and the FAIR Evaluation Service are not due to the metrics themselves, but to their different ways of collecting the metadata to be evaluated. This applies to all automated tools and contributes to the difficulty of comparison.

Krans et al. [31] provide a detailed qualitative comparison of ten tools, both questionnaires and automated tools. They mostly focus on their prerequisites, the ease and effort to use them, the type and quality of the outputs. They also test them on two datasets and observe a large variability in the FAIRness scores obtained, thus showing the difficulty in interpreting the questions in questionnaires and the differences in the implementations of the principles for automated tools. Candela et al. [10] provide an overview of twenty FAIR assessment tools, analyzing distinguishing features such as target, adaptability, methodology. They document the divergence between declared intents of metrics and what is actually assessed. Our work does not focus on the relevance of the implementations with respect to the declared intents. The authors also use a notion of coverage which shares some intuitions with our notions of coverage and granularity that we formally define. They also document co-occurrences between principles, where a co-occurrence means that "assessment metrics on a specific principle are also considered to be about many diverse principles". This is captured by our notion of dependency between implementations. In addition, we show and quantify the influence of such dependencies on the scores and on the way the measure behaves.

Some tools have also been compared in the context of domain-specific FAIRness assessment [23]. In particular, they compare the overall FAIRness score obtained on some datasets by the FAIR Evaluation Service, F-UJI, FAIRshake [11], and a self-assessment based on the FAIR data maturity model. They observe that the tools obtain scores close enough to consider that they give similar levels of FAIRness, especially if of the same kind (questionnaire or automated tool).

Recently, Moser et al. [37] propose a brief comparison of the FAIRness measures rather than the tools based on them. They focus on the FAIR Maturity Indicators [50], the FAIR Data Maturity Model [7], FAIRsFAIR metrics used in F-UJI [16] and FAIR metrics for EOSC [24]. They compare their numbers of indicators, and their structures: some metrics define indicators for intermediate principles (A1 and R1) while others do not. They also highlight that some of them give different importance to their indicators.

---

[1] With the metric collection: "All Maturity Indicator Tests as of May 8, 2019".

The main objective of our proposal is closer to this latter work, while we aim to push further the comparison of the scores and of the importance of each element in the measures. This is done by proposing a generic framework to model measures and formal definitions of characteristic quantities, that enable highlighting their salient traits. This article follows up the work presented in some of our publications. The starting point is described in [3–5], and focuses on accountability and the definition of a formal framework. However, all the definitions and sometimes the intuitions have evolved. This article does not propose a platform as in [35]. Compared to [32], the definition of the framework is more flexible, enabling the representation and analysis of the measures independently of the FAIR principles. Greater emphasis is placed on the notion of dependencies between implementations. We illustrate a methodology to identify them for the measure of FAIR-Checker and propose a definition of the impact that takes them into account. This article also includes a running toy example to illustrate the different notions. In order to be able to compare measures, we introduce the notion of alignment with a reference measure and the beginning of a small operator algebra. Finally, the experiments about FAIR-Checker that consider the dependencies are new, as the set of identified dependencies has been completed.

3   UReFM: A Representation Framework for Measures

With UReFM[2], our aim is to define a simple unifying model expressive enough to represent as many measures as possible, with some characteristic quantities.

Concerning the representation, we focus on the importance they give to the principles and sub-principles. In this view, our analysis of FAIRness measures enables to identify three notions to describe their tree-like organization: principles, indicators and implementations. As the ways scores are computed vary a lot, we use a generic representation of score computation. Hence, we propose to represent a measure as a tuple:

$$\mathcal{M} = (V, E, \text{Root}, \Diamond, w, D)$$

where elements (V, E, Root) refer to the structure, elements $(\Diamond, w)$ refer to the score computation, while D, possibly empty, is the set of the dependencies between implementations. These three constitutive parts are detailed in the next three subsections.

3.1   A Tree-Based Organization

The measures we consider rely on a hierarchy of principles (for example the FAIR hierarchy), the root of which is some node Root, for example FAIR, FAIR-Checker, OFAIRe... Here, the term *principle* is used in a broad sense, i.e. it refers to both principles and sub-principles. Given some measure $\mathcal{M}$, its set of principles is $P(\mathcal{M})$ and its edges, $E_P(\mathcal{M})$.

---
[2] UReFM: Unified Representation Framework for Measures.

Then, these principles are refined into several measurable criteria, expressed in natural language, which we call *indicators*. An indicator is named a "metric" in FAIR-Checker [20] and F-UJI [16], a "FAIRness assessment question" in O'FAIRe [1], a "maturity indicator test" in the FAIR Evaluation Service [50], or a "check" in FOOPS! [21]. Given $\mathcal{M}$, a measure of FAIRness, we denote $I(\mathcal{M})$ its set of indicators. Finally, indicators are associated to implementations, the set of which is noted $Imp(\mathcal{M})$. In the case of an automated tool, an *implementation* is an executable code run by a machine to provide a score for the evaluation of a digital resource.

Hence, the structure of a measure $\mathcal{M}$ is represented by a directed rooted tree (V, E, Root), simply adding indicators and implementations to the tree of principles, where:

- Root is the root of $\mathcal{M}$;
- V is the set of nodes of the structure, such that $V = P(\mathcal{M}) \cup I(\mathcal{M}) \cup Imp(\mathcal{M})$;
- E is the set of edges, where $E \subseteq E_P(\mathcal{M}) \cup (P(\mathcal{M}) \times I(\mathcal{M})) \cup (I(\mathcal{M}) \times Imp(\mathcal{M}))$;
- there can only be one implementation per indicator.

In this model, an implementation can be linked only to an indicator, which in turn can be linked only to a principle, any principle, not just to the leaves of the FAIR principles tree. This is intended to simplify the representation and to ease understanding and comparison of the measures.

An indicator, expressed in natural language, can be considered as a specification of the implementation. Indeed, the implementation of an indicator may vary according to the place of the indicator within the structure.

Both indicators and implementations may have several attributes, such as description, code, maximum score, etc. A dot notation is used to designate them (i.description, imp.code, i.URI...).

To manipulate a measure defined as above, we introduce the usual notions of children and of descendants. Let $n \in V$ be a node of the tree, then $\text{children}_{\mathcal{M}}(n)$ is the set of children of n in $\mathcal{M}$, and $\text{desc}_{\mathcal{M}}(n)$ is the set of descendants of n in $\mathcal{M}$. In addition, to denote the children and descendants of a node that are principles, we will use the notations: $\text{childPples}_{\mathcal{M}}(n)$ for $\text{children}_{\mathcal{M}}(n) \cap P(\mathcal{M})$ and $\text{descPples}_{\mathcal{M}}(n)$ for $\text{desc}_{\mathcal{M}}(n) \cap P(\mathcal{M})$. Such shortened notations are introduced as well for indicators: $\text{childInds}_{\mathcal{M}}(n)$, $\text{descInds}_{\mathcal{M}}(n)$; and implementations: $\text{childImpls}_{\mathcal{M}}(n)$, $\text{descImpls}_{\mathcal{M}}(n)$.

In this formalism, FAIR principles [49], which by definition have no indicators and no implementations, are represented by the structure $\mathcal{M}_{\text{FAIR}}$, illustrated Fig. 1.

$$\mathcal{M}_{\text{FAIR}} = (V_{\text{FAIR}}, E_{\text{FAIR}}, \text{FAIR}, \_, \_, \emptyset)$$

with:

- $V_{\text{FAIR}} = P(\mathcal{M}_{\text{FAIR}}) = \{\text{FAIR}, \text{F}, \text{F1}, \text{F2}, \text{F3}, \text{F4}, \text{A}, \text{A1}, \text{A1.1} \ldots\}$
- $E_{\text{FAIR}} = \{(\text{FAIR}, \text{F}), (\text{F}, \text{F1}), (\text{F}, \text{F2}) \ldots\}$

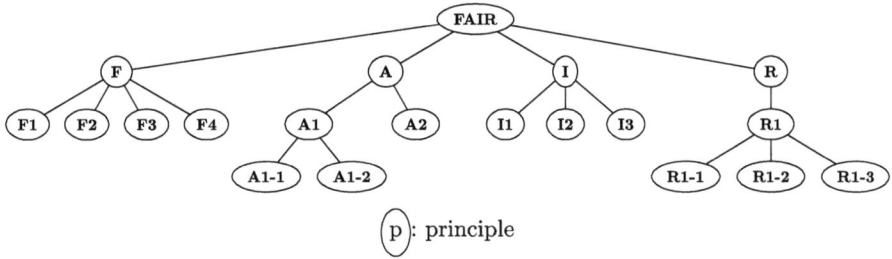

Fig. 1. Tree representation of FAIR Principles.

The various existing measures exhibit quite different characteristics from one another. So for the following, we use a toy example, named $\mathcal{M}_{\text{T-FAIR}}$, represented Fig. 2, which combines features from different measures to illustrate the concepts introduced. Similarly to mostly all measures, its number of indicators differs from one principle to another (one for F1 and F2, three for F4); some principles (F3, A1-1) are left without indicators (as, for example, in [18,20] for F3 and F4); some non-terminal principles (I) have indicator(s) (as for example in [19,21] for which an indicator is directly linked to A1); some principles (R2) have been added to the usual FAIR principles (as in [8], with additional principles R2 and R3).

3.2 Computing the Scores Through the Structure

In practice, considering existing measures, the ways scores are computed vary a lot. A choice has to be made to represent score computation. We favor an aggregated view of score computing, which is used by some measures, as in [1]. Intuitively, we consider that, given some digital resource d to evaluate, the score at a node is obtained by a weighted aggregation of the scores obtained by its children. The score of an indicator comes directly from executing its implementation imp for d. Thus we assume a family of evaluation functions, $\text{eval}_d : Imp(\mathcal{M}) \rightarrow \mathbb{R}^+$ such that $\text{eval}_d(imp)$ denotes the obtained score. In this view, we detail the elements $(\Diamond, w)$ of the representation of a measure.

- $\Diamond : \mathcal{P}(\mathbb{R}^+ \times \mathbb{R}^+) \rightarrow \mathbb{R}^+$ is an aggregation function, computing a new (aggregated) score from some pairs (*score*, *weight*). It can be either a weighted sum (noted SUM) or a weighted average (noted AVG). It returns 0 in case of an empty set.
  Note that function $\Diamond$ is assumed to be an increasing monotonic function: when its inputs increase, its result cannot decrease. To achieve this behavior, the weight has to be positive.
- $w : V \rightarrow \mathbb{R}^+$ is a weighting function. Given a node n, $w(n)$ represents the importance of n with respect to its siblings in the hierarchy[3].

---

[3] For simplicity, when a weight is not explicitly specified, it is assumed to be equal to one—for example, when a node is the single child of another one, such as for implementations.

For a given implementation imp, the minimum and maximum value that can be obtained at this level are respectively denoted imp.min-score and imp.max-score. For example, in F-UJI, i-R1-01MD.max-score = 4. Note that these two bounds have to be attainable. An overall score is obtained by the successive aggregation of the evaluations obtained at the implementation level. Then, the computation of the score obtained by some resource d can be expressed as follows.

$$\text{score}(\mathcal{M}, d) = \text{score}(\mathcal{M}, \text{eval}_d, \text{Root}) \qquad (1)$$

Function score, for a given node $n \in V$, is expressed for any function eval by:

$$\text{score}(\mathcal{M}, \text{eval}, n) = \begin{cases} \text{eval}(n) & \text{if } n \in Imp(\mathcal{M}) \\ \underset{n' \in \text{children}_{\mathcal{M}}(n)}{\Diamond} (w(n'), \text{score}(\mathcal{M}, \text{eval}, n')) & \text{otherwise} \end{cases} \qquad (2)$$

Let us illustrate the score computation. As shown by Fig. 2, the $\mathcal{M}_{\text{T-FAIR}}$ measure aggregates scores using an usual weighted average. Each indicator has the same weight (1) and the weight of a principle is the sum of the indicators' weights of its sub-tree, zero if there are none.

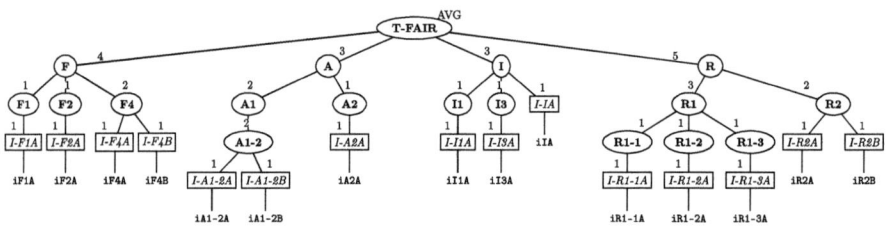

p : principle – id : indicator – imp: implementation – w: weight
for each implementation $imp.\texttt{min-score} = 0$ and $imp.\texttt{max-score} = 1$

Fig. 2. Running toy example.

According to this, the score of an entity d with respect to principle F is computed as follows (omitting the measure name in the function parameters):

$$\text{score}(\texttt{F}, d) = \frac{1 \times \text{score}(\texttt{F1}, d) + 1 \times \text{score}(\texttt{F2}, d) + 0 \times \text{score}(\texttt{F3}, d) + 2 \times \text{score}(\texttt{F4}, d)}{1+1+0+2}$$

$$= \frac{\text{score}(\texttt{F1}, d) + \text{score}(\texttt{F2}, d) + 2 \times \text{score}(\texttt{F4}, d)}{4}$$

$$= \frac{\frac{1 \times \text{score}(\texttt{I-F1A}, d)}{1} + \frac{1 \times \text{score}(\texttt{I-F2A}, d)}{1} + 2 \times \frac{1 \times \text{score}(\texttt{I-F4A}, d) + 1 \times \text{score}(\texttt{I-F4B}, d)}{1+1}}{4}$$

$$= \frac{\text{score}(\texttt{I-F1A}, d) + \text{score}(\texttt{I-F2A}, d) + \text{score}(\texttt{I-F4A}, d) + \text{score}(\texttt{I-F4B}, d)}{4}$$

$$= \frac{\text{eval}(\texttt{iF1A}, d) + \text{eval}(\texttt{iF2A}, d) + \text{eval}(\texttt{iF4A}, d) + \text{eval}(\texttt{iF4B}, d)}{4}$$

One can note that this way of positioning the weights corresponds exactly to averaging the score of the indicators, and we propose to represent in this way the real measures that achieve such an average.

3.3 (Non) Independence Between Implementations

Independence between implementations means that knowing the value obtained by a digital resource with some implementation gives no indication about the value obtained with the implementations used of the other indicators of the measure. However, this may not always be the case, leading us to introduce the term of dependency. For example, FAIR-Checker [18,20] explicitly exposes that there are some code delegations between implementations. To model the dependencies that occur within a measure, we introduce set D, a set of dependencies of the form:

$$\forall d, \text{eval}_d(imp) \; op_1 \; exp_1 \Rightarrow \text{eval}_d(imp') \; op_2 \; exp_2$$

where:

- $(imp, imp') \in Imp(\mathcal{M}) \times Imp(\mathcal{M})$,
- op_1 and op_2 are comparison operators belonging to $\{=, <, \leqslant, >, \geqslant\}$,
- exp_1 and exp_2 are mathematical expressions. More precisely:
  - exp_1 is a value or a variable.
  - exp_2 is a value if exp_1 is.
    Otherwise, exp_1 is a variable (v), and exp_2 defines a computation function that returns a value derived from v as in v, $v+1$ or $v/2$, etc.

For example, $\forall d, \text{eval}_d(imp) \geq 1 \Rightarrow \text{eval}_d(imp') \geq 1.2$ intuitively means that, for any resource, if executing imp results in a value above 1, one is warranted that implementation imp' results in a value above 1.2. This is what we call a dependency between imp and imp'.

To express that two implementations give the same results, we use the shortcut: $\forall d, \text{eval}_d(imp) = \text{eval}_d(imp')$.

This simple language could be extended in different ways, for example enabling a dependency to involve more than two implementations, but at this time, the current proposal enables to represent any of the concrete cases we have identified.

It is interesting to note that the notion of dependency does not play any role in the calculation of the score. In fact, in a situation where a digital object is evaluated by a measure, these mechanisms exist and operate whether or not one is consciously aware of them. Nonetheless, being aware of them enables a deeper understanding of how the measurement behaves. In particular, let us assume that the scores of some subset of implementations $S \subset Imp(\mathcal{M})$ is already known. We denote R-S the corresponding set of couples (imp, val). Being aware of the dependencies D enables an observer to determine whether the values obtained in R-S constrain the range of possible values (i.e. the lower and upper bounds) of the other implementations in $Imp(\mathcal{M})$. We denote:

- $\min_{R-S}^{\mathcal{M}}(imp')$, the minimal value that can be obtained by executing imp' given the set of known results R-S.
- $\max_{R-S}^{\mathcal{M}}(imp')$, the maximal value that can be obtained by executing imp' given the set of known results R-S.

As we will see in Subsect. 5.3, knowledge of dependencies is essential for accurate analysis of the relative importance of one principle within a measure. It is therefore essential for understanding the measurement and interpreting its results, as well as for comparing different measurements.

4 Modeling FAIRness Measures with UReFM

Many existing FAIRness measures can be represented with framework UReFM, some almost directly, while some points must be paid attention for others. In this section, we present the general process and discuss the main encountered issues. In doing so, we will mention several tools and FAIRness measures.

Among the tools referred to, two of them have the objective to assess any digital resources: FAIR-Checker [18,20], and FAIR Evaluation Service [2,50] (the indicators from the collection "All Maturity Indicator Tests as of May 8, 2019"). One tool focuses on datasets: F-UJI [16,17]. Two tools are designed specifically for ontologies: FOOPS! [19,21] and O'FAIRe[4] [1,38]. We also consider two online questionnaires with ARDC's questionnaire [6] and SATIFYD [14], allowing to make a self-assessment of a digital resource. We rely preferably on the online tools of the measures since they are more up-to-date and since it is usually easier to understand how the global score is computed.

In the following, we first consider the modeling of the structure, then representation of the score computation with UReFM. Finally, we propose different examples that illustrate the process.

4.1 Issues Related to the Modeling of the Structure

As long as the principles and sub-principles of a measure have a tree-like organization, with no "sub-indicators" (nor "sub-implementations"), the representation with UReFM is straightforward. All the nodes of the hierarchy are kept. If they exist, the dependencies between implementations must be expressed with the proposed language, which presents no difficulty as far as they are identified. Notice that identifying them is the difficult point since existing measures generally do not expose them. This requires analyzing all the implementations which may be a long and difficult task. Such work has for example been carried out in [10] with a different objective of analyzing discrepancies between implementations and indicators. The representation process described so far can be applied from hardly structured measures (for example questionnaires such as ARDC, which deal with principles F, A, I, R only) to measures that would include principles of the FAIR principles hierarchy and additional ones.

---

[4] To be precise, the represented version of O'FAIRe corresponds to the implementation deployed in bioportal. Note that, compared to the general definition of this measure, this implementation presents some specificities related to its particular context.

Notice that there may be measures where a same sub-principle has several parent nodes. In this case, we come back to a tree structure by duplicating the sub-structure as many times as necessary.

Let us now consider the cases where several levels of refinement are introduced to describe the indicators (or implementations). This is for example the case with Wilkinson et al. [50] or F-UJI [16]. Wilkinson et al. [50] propose a level of "maturity indicator", and then of "maturity indicator test" while F-UJI [16] has a level of "metrics" and then of "tests". To represent these measures with our framework, we have to reduce this hierarchy of indicators to a single level. This is done focusing on the ones highlighted by their measuring tools, the one for which a score is clearly given or the one to appearing clearly. Hence, we keep "maturity indicator test" and "metrics" respectively as the indicators. See Fig. 4 for F-UJI example.

Notice that, even though some of the particularities of some measures are not reproduced in their representation within the generic model, the proposed process maintains the characteristic elements necessary for understanding the importance they give to their principles. This is also the case for their score calculation which we explain in the following.

4.2 Scores Computation

In our framework, the scoring method is the combination of a weighted aggregation of the evaluations and scores obtained all along the tree structure, as explained in Subsect. 3.2. We seek to represent the scoring of existing measures in this way, ensuring that if a score is computed for a given (sub-)principle by the measure, the score computed through its representation is the same.

Existing measures with an automatic score computing, already have minimum and maximum values (min-score and max-score attribute) for implementations and indicators, an aggregation function $\Diamond$, and possibly a set D of dependencies between implementations. These elements are all kept as is in the representation of the measure. Only the weights remain to be defined.

Questionnaire approaches are again the simplest case. Usually scores are computed first on principles F, A, I and R, and then the global score is obtained as an average of these four scores. Accordingly, the aggregate function used in the representation is AVG (average), and the weights F, A, I and R are set to 1. Here, to identify possible dependencies, instead of the implementations, it is the asked questions which have to be analyzed.

Other measures use a deeper hierarchy and a higher number of indicators, but almost all of them compute the final score as the average the score obtained by the implementations, regardless of the hierarchy. Whatever the case, to represent a measure in our formalism, one has to determine the weight of each implementation and indicator, the weights on the other nodes and of course the function of aggregation. When this function or weights are made explicit, such as in O'FAIRe, where the weights of indicators and principles correspond to their "credits", they should be kept as is in the representation. Some approaches, for example the FAIR Evaluation Service, do not propose to compute any score.

This means that each user is supposed to choose how to use the different results obtained from implementations and indicators. If an overall score is required, again, it is up to the end-user to specify how this is to be obtained. In our formalism, this is equivalent to say that the choice of the function of aggregation and the weights is left to the end user. For example, if the measure computes the overall score summing the scores of indicators, its representation also uses a sum as the aggregation function and all the weights are set to 1. If the measure averages the scores of the indicators, its representation also uses an average as the aggregation function, the weights of the indicators are set to 1 and the weights of principles are set to the sum of their children weights.

4.3 Examples of Real-Life Measures Represented with UReFM

Our toy example T-FAIR (Fig. 2) is naturally represented with UReFM. In this section, the proposed representation process is illustrated with three real-life measures.

Questionnaire Approaches. Currently, only ARDC's tool [6] and SATIFYD [14] have been represented. Their structure only involves four of the FAIR Principles: F, A, I and R. When the computation of a global score is not part of their proposal, we propose an usual average with equal weights. And finally, since there is no implementation, we assume independence (D is empty). The representation of ARDC [6] is illustrated Fig. 3.

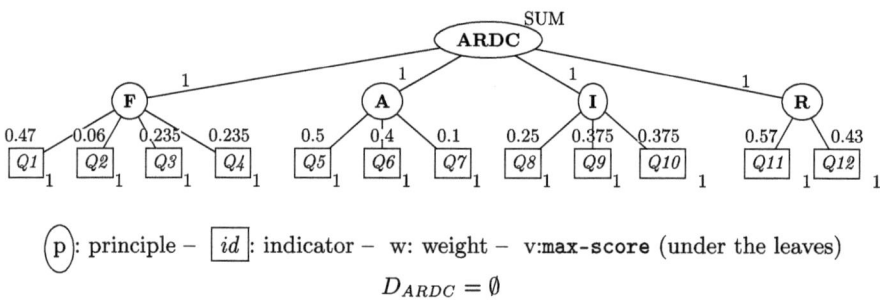

Fig. 3. Representation of ARDC [6] questionnaire.

F-UJI computes the score using an unweighted sum. Its representation is illustrated in Fig. 4. Notice that in F-UJI, value of max-score varies from 1 to 4, thus assigning different importance to indicators. Dependencies between implementations, if any, are not taken into account.

FAIR-Checker computes the score of a resource by averaging the values obtained from the indicators. Accordingly, in the representation, the weight of each indicator is set to 1 and an the function of aggregation is "average". To

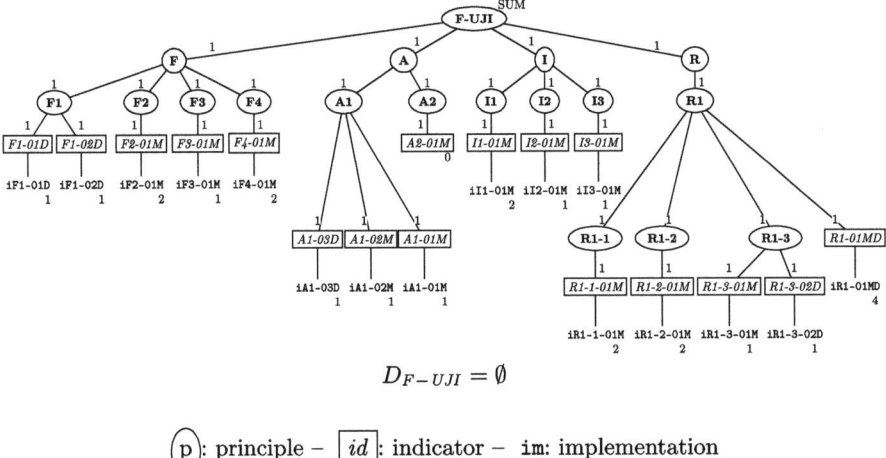

Fig. 4. Representation of F-UJI [7].

stick to the spirit and get the same score result, the weight of a node is assigned the sum of the weights of its children. This is illustrated in Fig. 5. FAIR-Checker computes the global score as a mean score, represented as a percentage, which is aligned with our proposed approach.

We in addition use FAIR-Checker to illustrate the analysis of possible dependencies between its implementations. First, FAIR-Checker directly exposes some dependencies within the code through delegations. According to these, implementations iF2A[8] and iI1 always output the same score. The same happens for the linked implementations iF2B, iI2 and iR1.3. According to our proposal, this can be represented by the following dependencies:

- $dep_1 : \forall d, \text{eval}_d(\text{iF2A}) = \text{eval}_d(\text{iI1})$
- $dep_2 : \forall d, \text{eval}_d(\text{iF2B}) = \text{eval}_d(\text{iI2})$
- $dep_3 : \forall d, \text{eval}_d(\text{iF2B}) = \text{eval}_d(\text{iR1.3})$

The links between iI2 and iR1.3 are obtained by transitivity.

Exposing code delegation is a good programming practice and should be encouraged. Unfortunately, dependencies are not exclusively the result of code delegation. Further analysis of the implementations is required.

The use of a formal code analysis method is not always easy nor possible, and a manual analysis may be quite tedious and difficult. For semantic web measures evaluating RDF triples, another way is possible. By inspecting FAIR-Checker's implementation, it is possible to identify minimal sets of triples benefiting the

---

[8] The code associated to implementation iF2A is accessible at url https://github.com/IFB-ElixirFr/FAIR-checker/blob/master/metrics/F2A_Impl.py. For the URL of other FAIR-Checker implementations, just replace F2A by their names.

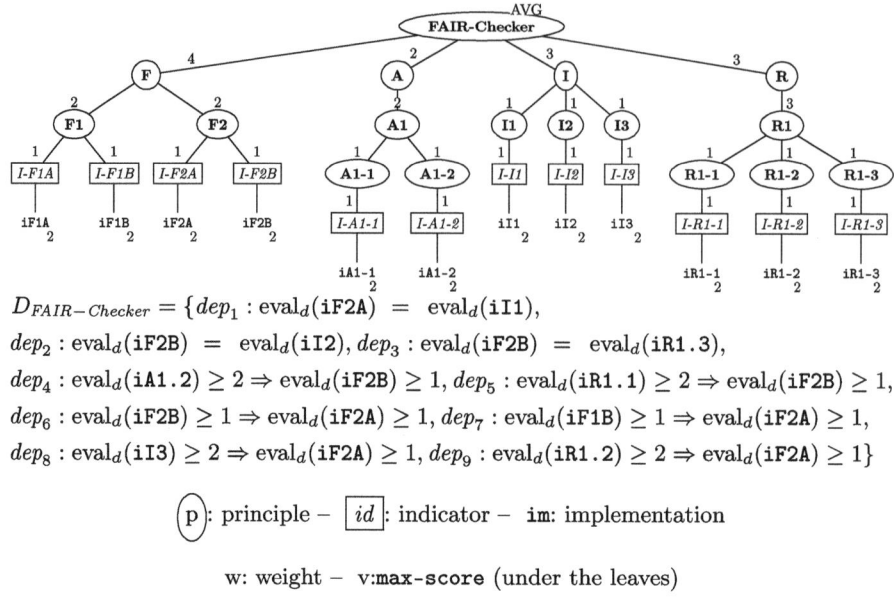

$D_{FAIR-Checker} = \{dep_1 : \text{eval}_d(\text{iF2A}) = \text{eval}_d(\text{iI1}),$
$dep_2 : \text{eval}_d(\text{iF2B}) = \text{eval}_d(\text{iI2}), dep_3 : \text{eval}_d(\text{iF2B}) = \text{eval}_d(\text{iR1.3}),$
$dep_4 : \text{eval}_d(\text{iA1.2}) \geq 2 \Rightarrow \text{eval}_d(\text{iF2B}) \geq 1, dep_5 : \text{eval}_d(\text{iR1.1}) \geq 2 \Rightarrow \text{eval}_d(\text{iF2B}) \geq 1,$
$dep_6 : \text{eval}_d(\text{iF2B}) \geq 1 \Rightarrow \text{eval}_d(\text{iF2A}) \geq 1, dep_7 : \text{eval}_d(\text{iF1B}) \geq 1 \Rightarrow \text{eval}_d(\text{iF2A}) \geq 1,$
$dep_8 : \text{eval}_d(\text{iI3}) \geq 2 \Rightarrow \text{eval}_d(\text{iF2A}) \geq 1, dep_9 : \text{eval}_d(\text{iR1.2}) \geq 2 \Rightarrow \text{eval}_d(\text{iF2A}) \geq 1\}$

(p): principle – $\boxed{id}$: indicator – im: implementation

w: weight – v:max-score (under the leaves)

Fig. 5. Representation of FAIR-Checker.

global evaluation. From them, detecting dependencies means looking for semantic inclusions between these sets. An automatic method is even possible, but outside the scope of this article. Table 4, in the Appendix at the end of this article, contains the different minimal patterns we have identified from the implementation. Some discussions are also provided about how to obtain dependencies from these elements. The graph of all identified dependencies is represented Fig. 6, quantification of d being omitted. The bold arrows represent the dependencies defined by code delegation. All the other arrows were identified using the minimal patterns based method.

The examples described in this section highlight the flexibility of the proposed framework. Notice that representing the measures with UReFM may change the way the score computing is viewed, but it does not add sub-principles nor suppress existing ones. For example, in Fig. 5 related to FAIR-Checker, sub-principles F3, F4 and A2 do not appear.

5 Characteristic Quantities for Measure Analysis

When considering a measure, one often wants to know what its intrinsic salient traits are, as for example: Are all the principles of the measure equally covered by indicators? What proportion of the score obtained by some digital resource is due to a given principle or sub-principle? To help answer some of these questions, we formally define three characteristic quantities of a measure: coverage rate, granularity and impact.

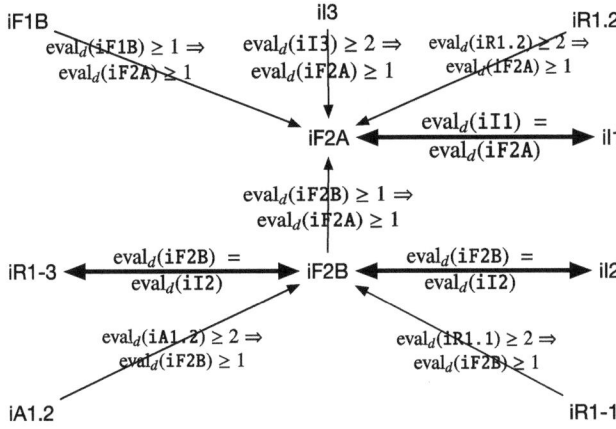

Fig. 6. FAIR-Checker graph of dependencies.

5.1 Coverage Rate

The role of the *coverage rate* is to measure to what extent the indicators of a given measure cover its principles. Several definitions may apply. The simplest would be the proportion of leaves of the principles tree that have an indicator. However, this does not take into account the specificity of all the measures, since for example, some of them have an indicator on A1 but neither on A1.1 nor A1.2. Such measures should have a higher coverage rate than others not covering any of these three principles. Therefore, one could consider all principles and calculate the proportion of them having an indicator in their descendants. This would correct the previous problem but it may lead to significantly high scores. For instance, with only one indicator on the principle A1.1, the coverage rate is of 20%.

Hence, we propose a coverage rate that takes into account the hierarchical aspect, where each sibling in the tree counts the same in the calculation of the coverage rate. In particular, considering the toy measure $\mathcal{M}_{\text{T-FAIR}}$, each of the four main principles counts as 25%.

Formally, let $\mathcal{M}$ be a measure, the *coverage rate* computed following the hierarchy is defined recursively by:

$$\text{cover}(\mathcal{M}) = \text{cover}(\mathcal{M}, \text{Root}) \tag{3}$$

where, for a given node $n \in P(\mathcal{M})$:

$$\text{cover}(\mathcal{M}, n) = \begin{cases} 0 & \text{if children}_{\mathcal{M}}(n) = \emptyset \\ \dfrac{\sum_{n' \in \text{childPples}(\mathcal{M},n)} (\text{cover}(\mathcal{M}, n')) + \text{hasInd}(\mathcal{M}, n)}{|\text{childPples}(\mathcal{M}, n)| + \text{hasInd}(\mathcal{M}, n)} & \text{else} \end{cases} \tag{4}$$

with

- hasInd$(\mathcal{M}, n) = \begin{cases} 1 & \text{if children}_\mathcal{M}(n) \cap I(\mathcal{M}) \neq \emptyset \\ 0 & \text{else} \end{cases}$
- childPples$(\mathcal{M}, n)$ = children$_\mathcal{M}(n) \cap P(\mathcal{M})$

Let us consider $\mathcal{M}_{\text{T-FAIR}}$ as an example. The coverage rate of a terminal principle is 1 if it has some indicator, like F1, F2, F4, A1-2, A2, I1, I3..., and 0 otherwise. Considering a non-terminal principles such as I, cover(T-FAIR, I) = $\frac{(1+1)+1}{2+1} = 1$. The global coverage score of $\mathcal{M}_{\text{T-FAIR}}$, noted cover(T-FAIR, T-FAIR), or simply cover(T-FAIR), is also equal to 1. Note that these values will evolve when the measure is aligned with the FAIR principles, as shown in Subsect. 6.2.

Several variations can be considered for this definition. First, to deal with automated measures, instead of focusing on indicators, considering implementations instead may be more indicative. This can be done replacing function 'hasInd' by 'hasImpl'. This adaptation would produce different results only if at least one indicator is not implemented. Such situations are present in existing FAIR measures and it is quite natural. For instance, an indicator may prove to be unimplementable. A second variation could be to pay attention to weights. Indeed, for a principle p with two sub-principles, p1 weighted 1 and covered, and p2 weighted 3 and not covered, one could consider that it is more intuitive p to be covered at 1/4 (instead of 1/2). For the current work, we have opted for the weight-neutral coverage to ease the comparison of measures.

5.2 Granularity

Granularity complements the previous definition. Intuitively, granularity assesses how much the indicators contribute to the analysis in complement to the structure of principles. In other terms, gran($\mathcal{M}$) evaluates the extent to which the indicators detail the principles embedded by the measure. gran($\mathcal{M}$, p) does the same for the sub-structure rooted by the principle p. Given a (sub)structure, higher its granularity is, deeper exploration of the principles and finer-grained analysis of the resources are. Practically, we quantify granularity of a node n as the ratio of the number of indicators that are descendants of n by the number of principles that are descendants of n (including n) and have at least a child which is an indicator.

$$\text{gran}(\mathcal{M}) = \text{gran}(\mathcal{M}, \text{Root}) \quad (5)$$

where, for $n \in P(\mathcal{M})$:

$$\text{gran}(\mathcal{M}, n) = \frac{|\text{descInds}_\mathcal{M}(n)|}{|\{p \mid p \in (\{n\} \cup \text{descPples}_\mathcal{M}(n)) \text{ and childInds}_\mathcal{M}(p) \neq \emptyset\}|} \quad (6)$$

As an example, let us compute the granularity of principles F and A in T-FAIR. Principle F has four indicators as descendants, while it has three descendant principles with indicators. Hence gran(T-FAIR, F) = 4/3, while gran(T-FAIR, A) = 3/2. This latter result is obtained because there only are two descendant principles of A which have indicators.

Similarly to the coverage rate, granularity is based on the indicators specified by the measure and can be easily adapted to account for implementations. The results provided by these two definitions diverge in the same scenarios as those presented for coverage rate.

5.3 Impact

By defining the impact of a node, we aim at quantifying how much this node contributes to the overall score of any given digital resource. For example, identifying principles with the strongest influence on the overall score may help a data provider to achieve a good score with minimum effort. Another potential use is to determine whether, from the point of view of a given measure, all the principles are considered equally important or, if not, what their relative importance are. In this section, we focus on the impact of principles, but the definitions are easily extendable to indicators and implementations.

Intuitively, the impact of a node n could be computed as a ratio, such as :

$$\text{impact}(\mathcal{M}, n) = \frac{BScore_{\mathcal{M},n}}{BScore_{\mathcal{M}}} \qquad (7)$$

where :

- $BScore_{\mathcal{M},n}$ represents the highest score that n can achieve by itself only. It can be obtained when "activating" the descendants of n only.
- $BScore_{\mathcal{M}}$ represents the highest score that can be obtained with the measure.

As the framework assumes that the aggregation function (weighted sum or average) is monotonically increasing, $BScore_{\mathcal{M},n}$ can be computed by aggregating the highest scores that can be obtained at each of the implementations that are in its scope (its descendants), while putting all the others at their minimal obtainable score. However, depending on the characteristics of the measure, computing these highest scores may not be trivial. In particular, it happens when dependencies exist between implementations. Indeed, it may be the case that using the highest value for an implementation prevents another one from reaching its minimal or maximal score. In a first time, we consider the case where all the implementations are independent. We consider the presence of dependencies in a second time. In every case, our goal is to define $BScore_{\mathcal{M},n}$ and $BScore_{\mathcal{M}}$, the impact being defined by the ratio of Eq. 7.

With Independent Implementations. The simplest case is when the scores that can be obtained at the implementations are normalized within a given interval $[0, max - score]$. Then, the impact of a node is obtained simply by considering the weights. The importance of a node n for its predecessor is obtained by: $\frac{w(n)}{\sum_{c \in \text{children}(\text{pred}(n))} w(c)}$. The impact of a node is obtained by multiplying these ratios all along the path between the node and the root.

When values `min-score` and `max-score` vary from an implementation to another, we need to express that the executions of all the implementations that are descendant of n are assumed at their maximum value while all others are set to their min-score value. This is formalized by function $best_n$ over implementations:

$$best_n(imp) = \begin{cases} \text{imp.max-score} & \text{if } imp \in \text{desc}(n) \\ \text{imp.min-score} & \text{otherwise} \end{cases} \quad (8)$$

where $imp \in Imp$. Then,

$$BScore_{\mathcal{M},n} = score(\mathcal{M}, best_n, \text{Root}) \quad (9)$$

and

$$BScore_{\mathcal{M}} = score(\mathcal{M}, best_{\text{Root}}, \text{Root}) \quad (10)$$

Notice that computing $score(\mathcal{M}, best_{\text{Root}}, \text{Root})$ is as simple as computing the overall score when all the implementations are set to their max-score value. As an example, let us consider the toy example measure T-FAIR, assuming no dependency between implementations. In that case, impact(T-FAIR, F) = $(4/15)/(15/15) = 4/15 \approx 0.2667$.

With Dependencies Between Implementations. Dependencies as presented in Subsect. 3.3 are expressed as a set D of rules specifying that the value obtained when executing an implementation may influence the value obtained with another one. In particular, there may be dependencies that limit the maximum or minimum score obtainable for an implementation. Notice that if no dependency limits the minimum nor the maximum value for any implementation, Eqs. 8, 9 and 10 still apply to compute $BScore_{\mathcal{M},n}$ and $BScore_{\mathcal{M}}$. According to our analysis, there are two possible cases.

Case 1: Only the minimum value of some implementations is limited by dependencies, such as in:

$$\text{dep}_{\text{F4B-I3A}} = \forall d, \text{eval}_d(\text{iF4B}) \geq 0.5 \Rightarrow \text{eval}_d(\text{iI3A}) \geq 0.5$$

In such a case, the definition of $BScore_{\mathcal{M}}$ remains unchanged. This is because all the implementations under Root are assigned their max-score value.

As for $BScore_{\mathcal{M},n}$, the implementations that are descendant of n are still assigned their max-score value. For each of the others, the minimum obtainable value given the dependencies must be computed. Consequently, a new function reflecting this specificity has to be introduced:

$$best_n^D(imp) = \begin{cases} imp.\text{max-score} & \text{if } imp \in \text{desc}(n) \\ \min_{\text{R-S}}^{\mathcal{M}}(imp) & \text{otherwise} \end{cases} \quad (11)$$

where $imp \in Imp$ and R-S $= \{(imp', imp'.\text{max-score}) : imp' \in \text{desc}(n) \cap Imp\}$. Then,

$$BScore_{\mathcal{M},n} = score(\mathcal{M}, best_n^D, \text{Root}) \quad (12)$$

and $BScore_{\mathcal{M}}$ is still given by Eq. 10.

Notice that when set D is empty, the previous equations compute the same impact as with Eqs. 8, 9 and 10.

As an example let us consider the impact of F on T-FAIR assuming $\text{dep}_{\text{F4B-I3A}}$ as the only dependency, i.e. $D = \{\text{dep}_{\text{F4B-I3A}}\}$. Computing $BScore_\mathcal{M}$ is done simply using the score function assuming that all the implementations are up to their maximum value (1). Then, as in the previous example, $score(\mathcal{M}, best_{Root}, \text{T-FAIR}) = \frac{4+3+3+5}{15} = 1$.

For computing of $BScore_{\mathcal{M},\text{F}}$, implementations which are descendant of F are set to their max-score value which is equal to 1. All those which are not descendant of F, except iI3A, are set to their min-score value which is 0, and iI3A is set to 0.5 because of dependency $\text{dep}_{\text{F4B-I3A}}$. In other words:

- $best^D_F(\text{iF1}) = 1$; $best^D_F(\text{iF2}) = 1$; $best^D_F(\text{iF4}) = 1$.
- $best^D_F(\text{iA1-2A}) = 0$; $best^D_F(\text{iA1-2B}) = 0$; $best^D_F(\text{iA2A}) = 0$; etc.
- $best^D_F(\text{iI3A}) = 0.5$

Then, $score(\mathcal{M}, best^D_F, \text{F}) = 1$, $score(\mathcal{M}, best^D_F, \text{A}) = 0$; $score(\mathcal{M}, best^D_F, \text{I}) = \frac{1\times0+1\times0.5+1\times0}{1+1+1} = 0.167$; and $score(\mathcal{M}, best^D_F, \text{R}) = 0$. So, $BScore_{\mathcal{M},\text{F}} = \frac{4\times1+3\times0+3\times0.167+5\times0}{4+3+3+5} = 0.3$.

So, when considering only the dependency $\text{dep}_{\text{F4B-I3A}}$, the impact of F in T-FAIR is: $\text{impact}(\text{T-FAIR}, \text{F}) = 0.3$, which is more that what was obtained without considering this dependency.

Case 2: The maximum value of some implementations are limited by dependencies such as in:

$$\text{dep}_{\text{F2A-I1A}} = \forall d, \text{eval}\, d(\text{iF2A}) \geq 0.5 \Rightarrow \text{eval}_d(\text{iI1A}) \leq 0.5$$

Clearly, if iF2A reaches its theoretical maximum (1), then iI1A cannot exceed 0.5, and thus cannot reach its own theoretical maximum.

In the presence of such dependencies, increasing the value obtained by an implementation may decrease another one. Thus $BScore_{\mathcal{M},n}$ can no longer by obtained by assigning their max-score value to the implementations that are descendants of node n. The problem becomes much more complex, needing the use of a solver. Fortunately, no such dependency has been identified in existing FAIRness measures. In the absence of any obvious need, this problem is out of the scope of this paper.

6 Alignment with the FAIR Principles

Representing different measures within a unifying framework is an important step towards enabling their comparison. However, even when measures are represented in the same framework, it remains difficult to compare those that assess fundamentally different aspects. For example, how can one compare a measure that evaluates code quality with one that focuses on performance? For a meaningful comparison, it is preferable to conduct it against a clearly defined reference. In our case, since our goal is to compare FAIRness measures, the FAIR principles themselves [49] serve as this reference structure, illustrated in Fig. 1.

As a consequence, we propose to align the measures with the FAIR principles, $\mathcal{M}_{\text{FAIR}}$. More precisely, for a given measure already represented in our formalism, we aim to derive an equivalent representation that uses all the FAIR principles and only those ones. This means that missing principles must be added, while those which are not part of the FAIR principles must be removed. Considering our toy example T-FAIR, Fig. 2, this implies that principle F3 which is missing must be added, while principle R2 must be removed. This new, aligned, measure has to be equivalent to the previous one and so to the measure itself. By equivalent, we mean the metrics' behavior on their common principles has to be exactly the same (they agree on the score of any digital resource), including the root.

6.1 Alignment Operators

Beyond the intuition previously explained, it is possible to define several operators to manipulate measures. In particular, an alignment with the FAIR principles structure can be achieved thanks to two operators: *addPple* and *dropPple*.

- *addPple* takes four parameters: the structure to modify, the name of the added principle, the full path to the node to which the added principle has to be connected, the relative weight of the added principle with respect to its siblings.
  For example, $addPple(\mathcal{M}_{\text{T-FAIR}}, \text{F3}, /\text{T-FAIR/F}/, 0)$ adds sub-principle F3 to the existing principle F of measure $\mathcal{M}_{\text{T-FAIR}}$ with the weight 0. Indeed, since the aggregation function is a weighted average, a zero weight fully neutralizes the principle: it is now present, but it plays no role. This is particularly useful when comparing multiple FAIR assessment measures with incomplete coverage of principles. Note it would be the same for a weighted sum.
- *dropPple* takes only two parameters: the structure to modify and the full path to the node to be removed. The nodes which were connected to the deleted principle are reconnected to the parent principle and their weights are adapted such that their impact does not change. example, $dropPple(\mathcal{M}_{\text{T-FAIR}}, /\text{T-FAIR/R/R2})$ removes the sub-principle R2 from measure $\mathcal{M}_{\text{T-FAIR}}$. After the removal of /T-FAIR/R/R2, I-R2A and I-R2B are connected to /T-FAIR/R and, as the aggregation function is a weighted average, since their previous weights were both 1 and the weight of /T-FAIR/R/R2 was 2, their new weights are $2 * \frac{1}{1+1}$, i.e. 1. Again, this also works for a weighted sum[9].

These two operators can be combined into a single one that would browse the two structures to implement the alignment. This may be the start for an operator algebra, to create, modify and combine measures.

---

[9] When the operator does not know how to compute new weights (for example due to the function of aggregation), it sets them to Null (unknown value), letting the user handle this question.

6.2 Examples of Alignment

The following examples illustrate the process of alignment.

T-FAIR. The representation of $\mathcal{M}_{\text{T-FAIR}}$ aligned on FAIR-principles, noted $\mathcal{M}_{\text{T-FAIR}}^{\mathcal{M}_{\text{FAIR}}}$, is represented Fig. 7 as well as the sequence of operations to obtain it from the initial representation.

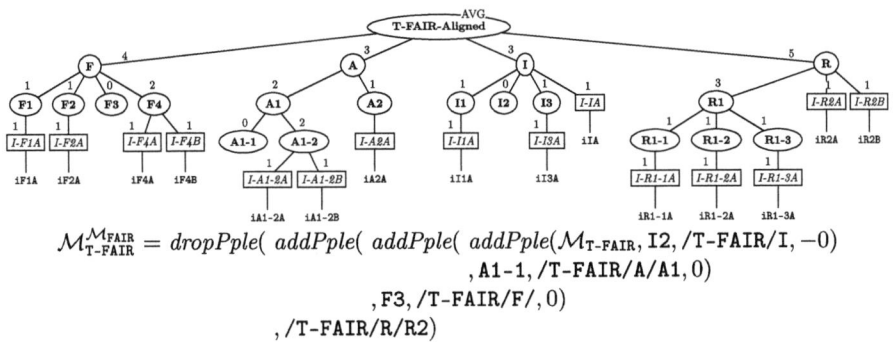

$$\mathcal{M}_{\text{T-FAIR}}^{\mathcal{M}_{\text{FAIR}}} = dropPple(\ addPple(\ addPple(\ addPple(\mathcal{M}_{\text{T-FAIR}}, \texttt{I2}, /\texttt{T-FAIR/I}, -0)$$
$$, \texttt{A1-1}, /\texttt{T-FAIR/A/A1}, 0)$$
$$, \texttt{F3}, /\texttt{T-FAIR/F/}, 0)$$
$$, /\texttt{T-FAIR/R/R2})$$

Fig. 7. T-FAIR representation aligned on FAIR-Principles.

Notice that, this alignment changes the coverage values. Now, added principles have a zero coverage, like F3, A1-1 and I2. This changes the coverage of their parents. For example, cover(T-FAIR-Aligned, I) = $\frac{(1+0+1)+1}{3+1}$ = 0.75. The global coverage rate of T-FAIR-Aligned is cover(T-FAIR-Aligned) = 0.8125 which clearly indicates that some principles are not covered.

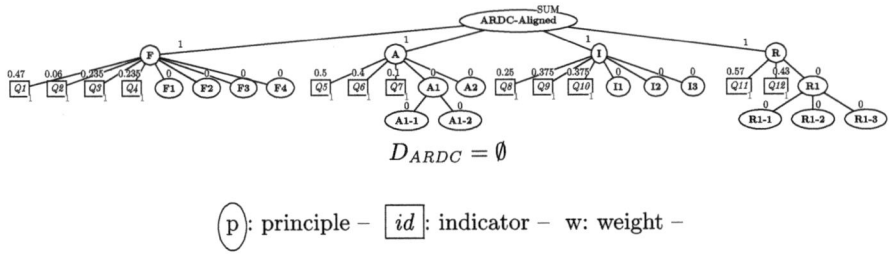

$$D_{ARDC} = \emptyset$$

(p): principle – [id]: indicator – w: weight –

Fig. 8. ARDC aligned on the FAIR Principles.

ARDC. Aligning questionnaires like ARDC is illustrated in Fig. 8. The only need is to add the missing principles, which have a coverage equal to 0, because they have no indicator.

FAIR-Checker can be also be aligned with operator *addPple* only. Indeed, only principles F3, F4 and A2 are missing. Hence they are added with a weight equal to zero. Their coverage is also equal to zero. The set of dependencies keeps unchanged. The alignment of FAIR-Checker is illustrated in Fig. 9.

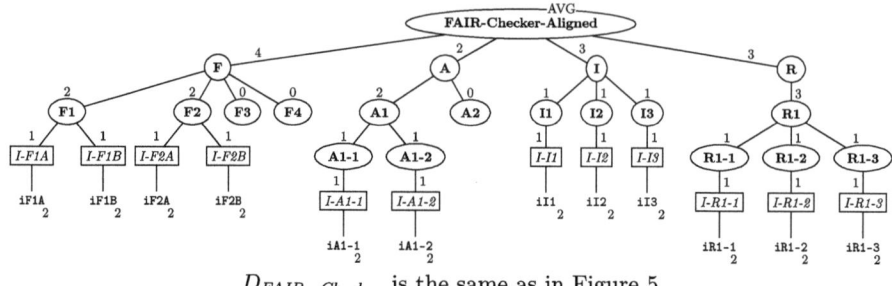

$D_{FAIR-Checker}$ is the same as in Figure 5

(p): principle – |id|: indicator – im: implementation
w: weight – v: max-score (under the leaves)

Fig. 9. FAIR-Checker aligned on the FAIR Principles.

F-UJI only lacks principles A1-1 and A1-2, which are added using the *addPple* operator. Thus, their weight is zero. In Subsect. 3.3 F-JUI has been represented with an empty set of dependencies. So it is after being aligned, as represented in Fig. 10.

Introducing the notion of alignment with a reference measure, with associated operators, enables to represent the different FAIRness measures aligned with the FAIR principles $\mathcal{M}_{FAIR}$ (Fig. 1). With the FAIR principles as a reference, it is possible to compare all the aligned measures within the framework, considering the characteristic quantities defined in Sect. 5. Such comparisons complement those that are conducted studying the scores obtained by different digital resources with the tools that implement the measures. These experimental studies are the purpose of the next section.

7 Experimental Results

Through these experiments, we first show that digital resources may obtain very different FAIR assessment scores when using different FAIR evaluation engines. Then, we show that our proposed framework can precisely document the coverage and the relative importance, for each tool, of both fine-grained FAIR indicators as well as global principles, thus providing insights for users and tool developers on possible evaluation biases. Additional material is available online[10].

---
[10] https://github.com/ICG4FAIR/ICG4FAIR.

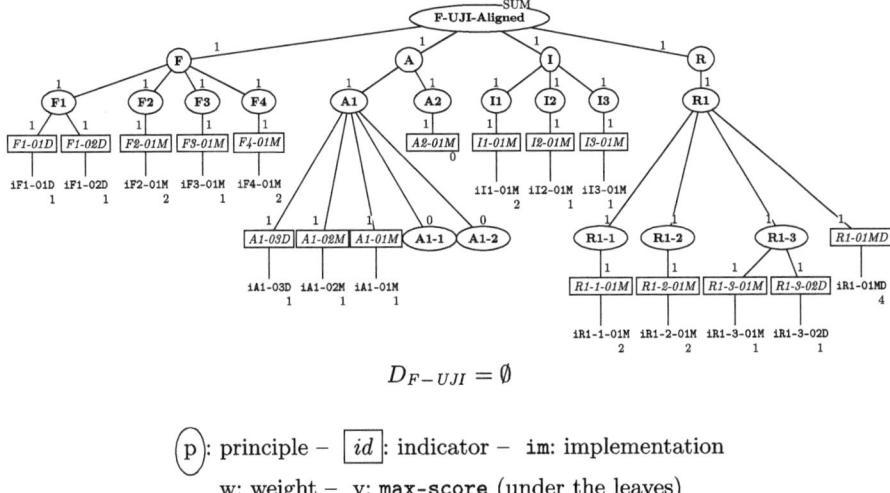

$D_{F-UJI} = \emptyset$

Ⓟ: principle – ⬚id⬚: indicator – im: implementation
w: weight – v: max-score (under the leaves)

Fig. 10. F-UJI aligned on the FAIR Principles.

In this section, the characteristic quantities, impact, coverage and granularity are computed with the FAIR principles as reference, using the aligned version of the measures, so that they can be compared more easily. However, for easier presentation, we keep the usual name of the measures, as for example F-UJI instead of F-UJI-Aligned.

All the impact values considered in Subsects. 7.1 and 7.2 are computed assuming that set D of dependencies is empty. In some way, this corresponds to the view that any user could have. Indeed, the existing dependencies are generally not exposed by the designers of measures and tools, and not everyone can identify them. In order to nevertheless illustrate the importance of dependencies, in Subsect. 7.3, we consider FAIR-Checker, showing that the impact values obtained with and without considering dependencies may greatly differ.

7.1 Do FAIR Assessment Engines Reach Consensus on FAIRness?

Table 1 reports FAIR assessments, performed the 3rd of March 2024, for a collection of 10 scientific digital resources with F-UJI and FAIR-Checker. In this selection, we aim at covering diverse domains with different types of resources such as datasets, ontologies, software, or training material. These resources are exposed on the web through diverse modalities such as institutional open data platforms, community specific registries (bioinformatics tools, machine learning models), e-learning platforms, legacy websites or raw metadata.

The F-UJI scores, expressed in percentages, have been manually collected from the tool's web interface [17]. The FAIR-Checker scores have been collected through the tool's API [18]. Since FAIR-Checker only provides fine-grained scores

Table 1. Multiple FAIR assessments, ranked by standard deviation.

Resource	F-UJI (%)	FAIR-Checker (%)	Std dev
Dataset (PANGAEA) [39]	91	91.70	0.49
Gene Ontology (OLS) [25]	18	16.70	0.92
Dataset (Harvard Dataverse) [27]	75	79.20	2.97
Dataset (Kaggle) [30]	60	70.80	7.64
Online course (Moodle) [36]	4	16.70	8.98
Dataset (Governmental platform) [26]	52	70.80	13.29
Dataset (WHO) [48]	27	50.00	16.26
Training material (TeSS) [44]	39	70.80	22.49
Bioinformatics tool (bio.tools) [9]	18	54.20	25.60
Dataset (RDF metadata) [41]	43	87.50	31.47

Table 2. Evaluating a bio.tools catalogue record.

	FAIR Score (%)	F (%)	A (%)	I (%)	R (%)
F-UJI	18.8	**35.7**	33	0	10
FAIR-Checker	54.2	**75**	50	**66.7**	16.7

as reported in Subsect. 4.3, we computed a global score as a percentage based on the maximum achievable score.

Table 1 shows a relatively good agreement between the two engines for the first 5 entries (50% of our collection) with a standard deviation lower than 10. The best agreement appears for very high or very low scores. However, the evaluations provided by F-UJI appear to be more fine-grained. The online course (Moodle) and Gene Ontology both obtain the FAIR-Checker minimal score (16.7%) but obtain different scores with F-UJI, 4% and 18% respectively, suggesting that F-UJI evaluation is more detailed. In addition, for the second half of our resource collection, the FAIR assessment scores begin to diverge with a standard deviation ranging from 13.29 to 31.47. Globally, we observe that FAIR-Checker provides higher scores compared to F-UJI.

For the last two entries the scores are very different with a standard deviation greater than 25. It is completely reasonable to wonder why the evaluation results are so different. Is it due to the way FAIR assessment engine retrieve metadata, as described in [45]? Is it due to the engine inner implementations? In the next paragraphs, we feed our model and compare FAIR-Checker and F-UJI, when evaluating a record of a bioinformatics tools catalogue.

Now we focus on how impact, granularity and coverage quantities can help in understanding divergent FAIR assessments. Table 2 reports very different results for the global FAIR assessment of the bio.tools record [9]. If we explore in more details each individual principle, we highlight that findability and interoperability scores greatly differ.

Table 3. Comparing F-UJI and FAIR-Checker (FC)

	Impact		Granularity		Coverage	
	F-UJI	FC	F-UJI	FC	F-UJI	FC
F	29.17	**33.33**	1.25	**2**	100	**50**
A	12.5	**16.67**	**2**	1	66.67	**50**
I	**16.67**	**25**	1	1	100	100
R	**41.67**	**25**	1.25	1	100	100

Table 3 shows how F-UJI and FAIR-Checker differ in terms of impact, granularity and coverage. We can see that reusability has a highest impact (41.67%) on the global assessment score compared to FAIR-Checker (25%). This could contribute to the explanation of the very low global FAIR assessment of the bio.tools record in F-UJI (Table 2) compared to FAIR-Checker. The findability of the bio.tools record is better scored in FAIR-Checker (75%) compared to F-UJI (35.7%). However, this result should be interpreted with caution due to a poor coverage of F principles in FAIR-Checker (50%), compared to F-UJI (100%). In addition, despite a low coverage, we show that this findability principle has still the higher impact (33.33%) in the global assessment, which is questionable. Regarding the interoperability, we observe 0% for F-UJI and 66.7% for FAIR-Checker. Both engines have the same granularity (1) for an 100% coverage, meaning that the two engines, by design, do not agree on the indicators for interoperability, or that their implementation greatly vary.

7.2 Comparison of Measures Based on the Characteristic Quantities

We illustrate the use of the characteristic quantities of the measures introduced in Sect. 5 to objectively highlight their salient features and some of their differences. All the seven measures considered in Sect. 4 have been represented in the framework the following the methodology explained in the same section. They also have been aligned with the FAIR principles as explained in Sect. 6.

Coverage Rates and Granularities. The coverage rate and granularity of all the principles, for all the seven measures are shown in Fig. 11 and Fig. 12 respectively. Notice that concerning O'FAIRe, the coverage rate and the granularity do not consider that some of its indicators are not (yet) implemented. This does not change the coverage value since all principles have at least one indicator implemented. However, the granularity is slightly overestimated.

We highlight some elements of the analysis that can be made. First, the coverage rate of root Root is rarely equal to 1. This means that the majority of the measures we are studying do not cover one or several principles. Only O'FAIRe covers all the principles. Questionnaires ARDC and SATIFYD have a low coverage rate because they only consider principles F, A, I and R. For

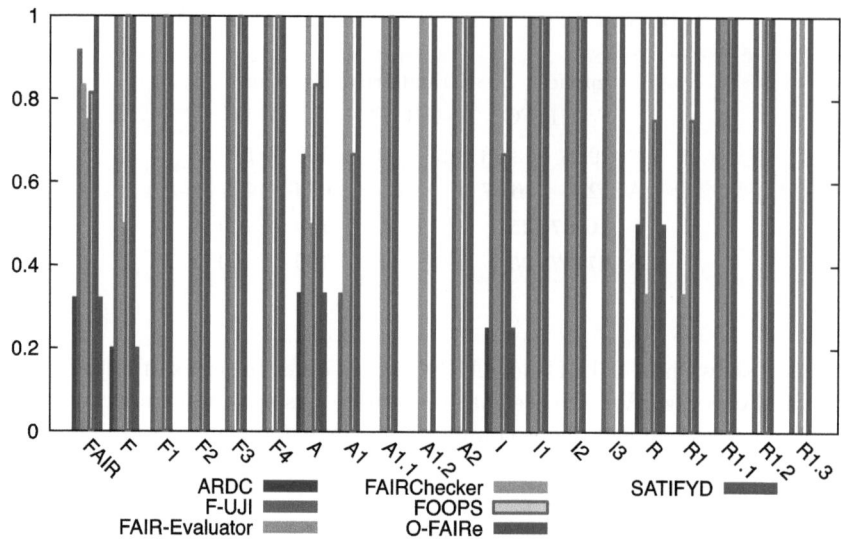

Fig. 11. Coverage rates.

FOOPS!, the coverage rate of A is 0.83. Its value is 0.66 for A1 whereas one would expect less because the value is 1 for A1.1 while A1.2 is not covered at all. This is because of the presence of an indicator directly related to A1. We also observe that some sub-principles are not covered by each measure. By design, R1.2 and R1.3 are not covered in the FAIR Evaluation Service as well as F3 and F4 for FAIR-Checker. In addition some principles such as A1.2 and R1.3 are covered by only a few number of tools, which questions on the technical feasibility of their implementation.

O'FAIRe gets the highest granularity at root FAIR, with high scores for I2 and R1.2 in particular. This is because it defines many indicators, may be to address the great variety of vocabularies and meta-data present within the semantic artifacts it usually assesses. As for FOOPS!, the high value of granularity of F1 shows an important care in providing indicators for this principle. Notice too that the granularity of R1 is higher than the granularities of R1.1, R1.2, R1.3. This is because several indicators have been directly attached to R1.

In fact, granularity and coverage rate are complementary and should sometimes be considered together before drawing a conclusion. For example, the promising granularity of F for ARDC could mean that F has been paid a lot of attention. However, the coverage rate of F for ARDC is quite low, meaning that there is no analysis according the sub-principles of F. Contrary to what one could expect, the analysis is not that precise.

Impacts. We then compare the different measures according to the impact, i.e. the importance they give to each principle. Figure 13 illustrates the impact of the four main principles, but we could compare the measures in detail according to

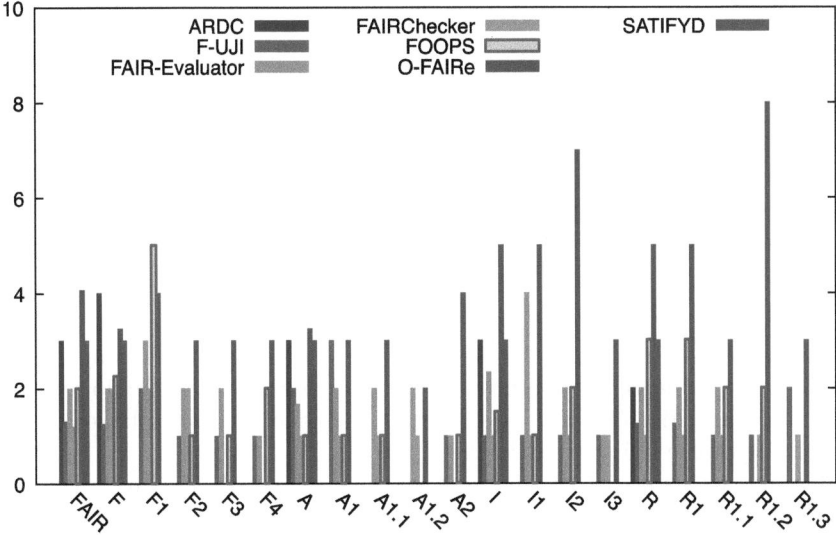

Fig. 12. Granularities

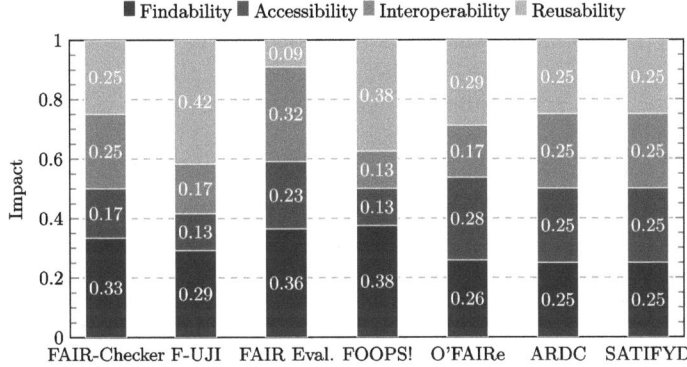

Fig. 13. Impact of the main FAIR Principles for different measures.

each sub-principles. Findability and Reusability are the most important principles in general. Reusability even counts as 42% of the maximum score for F-UJI. But unlike the other measures, it is of really low importance for FAIR Evaluation Service, which does not consider R1.2 and R1.3. Overall, these measures are not balanced, with a principle that is almost three times more important than another one for three of the five considered measures.

Apart from O'FAIRe, none of the measures really thinks about the importance to give to each indicator. This is understandable from the point of view of certain measures. For instance, Wilkinson *et al.* [50] insist on the fact that FAIRness measures should (only) act as an incentive to improve the FAIRness of digital resource, that "there is no intrinsic value in an evaluation score" and that we should not declare a resource FAIR or non-FAIR. However, even if the measures are only intended to push for improvement in the FAIRness of resources, it is a unfortunate that there is no order of priority. The SHARing Rewards and Credit (SHARC) Interest Group [15] and the Data Maturity Model Working Group [7], both from the Research Data Alliance, developed this idea that some criteria are more important than others by categorizing them as "useful", "important" or "essential".

7.3 Impact Values Under Dependencies

In this subsection, our aim is to highlight the changes obtained computing the impact when there are dependencies between implementations. To do so, we use FAIR-Checker (Fig. 5), which has several dependencies, as explained in Subsect. 4.3, with the full analysis in the Appendix.

The impact values of the different principles and sub-principles have been first computed assuming that set $D_{FAIR-Checker}$ is empty. Then, all these values have been re-computed considering the dependencies. Let us illustrate what it means for principle A1. According to the definition of the impact in the context of dependencies, as explained Subsect. 5.3, implementations iA1-1 and iA1-2 are set to 2, their maximum value, while the others take their minimum value. However, in the case of dependencies described in Fig. 6, this latter is not 0: A successful evaluation of iA1-2 implies that iF2B is always greater or equal to 1. Since there is no other influence, the minimum possible value of iF2B is 1. Then, iR1-3, iI2 and iF2A are also set to 1. So is iI1, due to the dependency with iF2A. This leads A1 to have an impact value equal to 0.2917 while it is only equal to 0.1667 without considering dependencies.

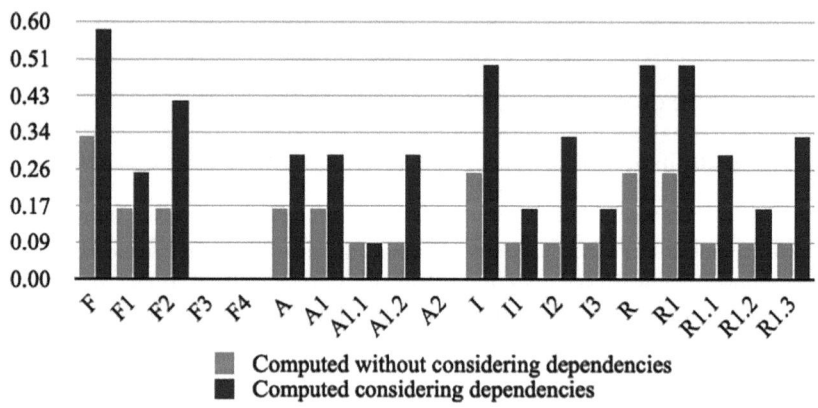

Fig. 14. Impact values of FAIR principles, according to FAIR-Checker.

Figure 14 shows all the impact values that have been computed in both cases. Except for A1.1, the impact values are higher when considering the dependencies. This is because these latter may require the minimal value of some implementations to be greater than 0, influencing the overall score. The most influencing ones are those resulting from code delegation. Indeed, when computing impact(F), implementations iI1 is set to iF2A.max-score, while iI2 and iR1-3 are set to iF2B.max-score. This further increases the value obtained. Hence, the success of all the implementations belonging to desc(F) ensures to obtain not 33% but 58% of the maximum possible score.

From an user point of view, such analysis is quite important, since it reveals that F is much more central to this measure than one might think at first glance. Thus, satisfying all the indicators under F should be the priority of an user seeking to get a good global score.

8 Conclusion and Perspectives

In this paper, we introduced UReFM, a formal generic framework. It enables a unified representation of measures and definition of three computable objective quantities. Together, the framework and the defined quantities facilitate both interpreting measures organized according to a hierarchy, such as FAIRness measures and comparing tools and possibly revealing evaluation biases. Our experiments show that our framework i) contributes to explain different scores obtained by the same digital artifacts using different assessment engines and ii) facilitates the setup of comparative studies of various FAIRness metrics. In the future, we intend to conduct larger scale experiments, considering more digital resources and more FAIR assessment tools.

In many implemented tools, the independence of implementations hypothesis is not satisfied [10]. Our framework enables modeling such dependencies, and the definition proposed for impact takes into account the consequences of their presence. To illustrate their influence, we used FAIR-Checker, analyzing the code of the implementations, to identify the several other dependencies, beyond those exposed as code delegation. The experiments clearly show that success for some well chosen implementations ensures a good score. Our study also underlines some difficulties. First, computing the impact when the dependencies limit the maximum values of some others becomes difficult, requiring at least a solver. Second, browsing the code of the implementations and identifying the dependencies is not easy. Clearly, the work conducted in [10] can help, as they identify co-occurrences in the implementations. In any case, it would be beneficial for designers to minimize dependencies whenever possible. Their presence significantly increases the complexity of both understanding the measure and interpreting its results. However, this may not be always possible. As a good practice, code delegations should be exposed (as done with FAIR-Checker), and dependencies too. By making these dependencies explicit in UReFM, we contribute to highlight potential biases for users when executing measures. This is also useful for developers to reveal the impact of technical choices on the relative importance given to some of the evaluated principles.

Impact, coverage and granularity clearly help analyzing measures. Their definition could be improved. For example, some of our choices induce some inaccuracies concerning granularity. Indeed, F-UJI divides its indicators into a new level of tests that would increase its granularity if considered. The proposed quantities could also be complemented by other ones. For instance, a coarse-grained scoring system may limit the ability to reflect subtle differences. We thus believe that considering the *discriminative power* of an implementation, a principle and a measure—their ability to assign many distinct scores to different digital objects—can provide valuable insight into their behavior and help comparison with other measures.

Our framework is generic enough to be considered for other domains, in which general principles or guidelines need to be evaluated. This is the case for instance with GDPR compliance [22], which can be interpreted differently depending on national regulatory policies. We could also imagine applications in the domain of trustworthy AI [33], or energy footprint [47].

As future works, we also intend to contribute to a service designed for a broad community that would allow (i) the description and sharing of measure specifications in a machine-readable way, and (ii) the reuse and combination of parts of these specifications through operators, to meet community-specific needs. This platform would also enable the linking of assessments to their formal specification thereby enhancing interpretability and mitigating the black-box effect of evaluations.

Acknowledgments. This work has been supported by project ANR DeKaloG (Decentralized Knowledge Graphs), ANR-19-CE23-0014, CE23 - Intelligence artificielle.

Disclosure of Interests. The authors have no competing interests to declare that are relevantto the content of this article.

Appendix: Dependencies between implementations in FAIR-Checker.

Table 4 summarizes the findings from systematically analyzing the FAIR-Checker code available on GitHub: https://github.com/IFB-ElixirFr/FAIR-checker/blob/master/metrics.

The "Minimal Pattern" column contains:

- whenever the implementation code corresponds to a triple search (SPARQL ASK query), a minimal set of triples required to pass the test. Possible variations indicated in the code are listed in the "Local Semantic Alternatives" column.
- in other cases, the specification provided in the code as a comment which is preserved as is.

A person familiar with programming languages can easily verify that these elements can be extracted simply reading the code.

Table 4. FAIR-Checker minimal patterns

Global semantic equivalences			
None			
Implem	minimal pattern	local semantic alternatives	score
iF1A	# Status code is OK, meaning the url is Unique.		2
iF1B	# At least one namespace from identifiers.org is used in metadata		1
	(?sF1B, dct:identifier, ?oF1B)	{dct:identifier, schema:identifier}	2
iF2A	# one triple in metadata (?sF2AW, ?pF2AW, ?oF2AW)		1
	(?sF2AS, dct:title, ?oF2AS)	{dct:title, dct:description, dcat:accessURL, dcat:downloadURL, dcat:endpointDescription, cat:endpointURL}	2
iF2B	# at least one used ontology class or property known in major ontology registries (OLS, BioPortal, LOV)		1
	# All classes and properties are known in major ontology registries (OLS, BioPortal, LOV))		2
iA1.1	# The resource uses HTTP protocol		2
iA1.2	(?sA1_2, odrl:hasPolicy, ?oA1_2)	{odrl:hasPolicy, dct:rights, dct:accessRights, dct:license, schema:license}	2
iI1	# one triple in metadata (?sF2AW, ?pF2AW, ?oF2AW)		1
	(?sI1, dct:title, ?oI1)	{dct:title, dct:description, dcat:accessURL, dcat:downloadURL, dcat:endpointDescription, cat:endpointURL}	2
iI2	# at least one used ontology class or property known in major ontology registries (OLS, BioPortal, LOV)		1
	# all classes and properties are known in major ontology registries (OLS, BioPortal, LOV))		2
iI3	# At least 3 different domains were found in metadata		2
iR1.1	(?sR1_1, dct:license, ?oR1_1)	{schema:license, dct:license, doap:license, dbpedia-owl:license, cc:license, xhv:license, sto:license, nie:license}	2
R1.2	(?sR1_2, dct:creator, ?oR1_2)	{prov:wasGeneratedBy, prov:wasDerivedFrom, prov:wasAttributedTo, prov:used, prov:wasInformedBy, prov:wasAssociatedWith, prov:startedAtTime, prov:endedAtTime, dct:hasVersion, dct:isVersionOf, dct:creator, dct:contributor, dct:publisher, pav:hasVersion, pav:version, pav:hasCurrentVersion, pav:createdBy, pav:authoredBy, pav:retrievedFrom, pav:importedFrom, pav:createdWith, pav:retrievedBy, pav:importedBy, pav:curatedBy, pav:createdAt, pav:previousVersion, schema:creator, schema:author, schema:publisher, schema:provider, schema:funder}	2
iR1.3	# at least one used class or property known in major ontology registries (OLS, BioPortal, LOV)		1
	# all classes and properties are known in major ontology registries (OLS, BioPortal, LOV))		2

From the elements presented in the previous table, the following dependencies are identified:

- $dep_1 : \forall d, \mathrm{eval}_d(\mathtt{iF2A}) = \mathrm{eval}_d(\mathtt{iI1})$: equivalence of implementations iF2A and iI1.
  In the table, the entries for these two implementations are identical: the same entry produces the same scores.
  This is directly due to the use of delegation in the program.
- $dep_2 : \forall d, \mathrm{eval}_d(\mathtt{iF2B}) = \mathrm{eval}_d(\mathtt{iI2})$ and
  $dep_3 : \forall d, \mathrm{eval}_d(\mathtt{iF2B}) = \mathrm{eval}_d(\mathtt{iR1.3})$: equivalence of implementations iF2B, iI2, and iR1.3.
  same as previous point.
- $dep_4 : \forall d, \mathrm{eval}_d(\mathtt{iA1.2}) \geq 2 \Rightarrow \mathrm{eval}_d(\mathtt{iF2B}) \geq 1$: if iA1.2 succeeds then the weak condition of F2B is satisfied.
  Indeed, iA1.2 tests the presence of a triple. If it is a success, at least one property is known in major ontologies registries since all alternatives are known in LOV ([34] provides tools to check this point).
- $dep_5 : \forall d, \mathrm{eval}_d(\mathtt{iR1.1}) \geq 2 \Rightarrow \mathrm{eval}_d(\mathtt{iF2B}) \geq 1$.
  Same as previous point.
- $dep_6 : \forall d, \mathrm{eval}_d(\mathtt{iF2B}) \geq 1 \Rightarrow \mathrm{eval}_d(\mathtt{iF2A}) \geq 1$.
  A success of iF2B with a score greater or equal to 1 indicates metadata contains at least one used ontology class or property known in major ontology registry. So, there is at least one triple in the metadata leading to the success of iF2A with a score greater or equal to 1.
- $dep_7 : \forall d, \mathrm{eval}_d(\mathtt{iF1B}) \geq 1 \Rightarrow \mathrm{eval}_d(\mathtt{iF2A}) \geq 1$.
  Same as previous point.
- $dep_8 : \forall d, \mathrm{eval}_d(\mathtt{iI3}) \geq 2 \Rightarrow \mathrm{eval}_d(\mathtt{iF2A}) \geq 1$.
  Same as dep_7.
- $dep_9 : \forall d, \mathrm{eval}_d(\mathtt{iR1.2}) \geq 2 \Rightarrow \mathrm{eval}_d(\mathtt{iF2A}) \geq 1$.
  A success of iR1.2 with a score greater or equal to 2 indicates metadata contains a triple of the form (_, p, _) with p being one semantic alternative. So, there is at least one triple in the metadata leading to the success of iF2A with a score greater or equal to 1.

Some other dependencies are obtained by deduction from those ones (transitivity). For example, $\forall d, \mathrm{eval}_d(\mathtt{iR1.1}) \geq 2 \Rightarrow \mathrm{eval}_d(\mathtt{iI1}) \geq 1$ is obtained combining dep_5, dep_6 and dep_1.

References

1. Amdouni, E., Bouazzouni, S., Jonquet, C.: O'faire makes you an offer: metadata-based automatic fairness assessment for ontologies and semantic resources. Int. J. Metadata Semant. Ontol. **16**(1), 16–46 (2022)
2. Fair evaluation service. https://w3id.org/AmIFAIR. Accessed: 18 June 2025
3. Andersen, J.: From transparency of knowledge graphs to a general framework for defining assessment measures. Ph.D. thesis, INSA Lyon, France (2024). https://tel.archives-ouvertes.fr/tel-04874933

4. Andersen, J., Cazalens, S., Lamarre, P.: Assessing knowledge graphs accountability. In: Pesquita, C., et al. (eds.) The Semantic Web: ESWC 2023 Satellite Events - Hersonissos, Crete, Greece, May 28 - June 1, 2023, Proceedings. LNCS, vol. 13998, pp. 37–42. Springer (2023). https://doi.org/10.1007/978-3-031-43458-7_7
5. Andersen, J., Cazalens, S., Lamarre, P., Maillot, P.: A framework to assess knowledge graphs accountability. In: IEEE International Conference on Web Intelligence and Intelligent Agent Technology, WI-IAT 2023, Venice, Italy, 26-29 October 2023, pp. 213–220. IEEE (2023). https://doi.org/10.1109/WI-IAT59888.2023.00034
6. Australian Research Data Commons (ARDC): FAIR self assessment tool: ARDC online questionnaire (2022). https://ardc.edu.au/resource/fair-data-self-assessment-tool/. Accessed 22 Apr 2024
7. Bahim, C., et al.: The fair data maturity model: an approach to harmonise fair assessments. Data Sci. J. **19**, 41–41 (2020)
8. Barker, M., et al.: Introducing the fair principles for research software. Sci. Data **9** (2022). https://api.semanticscholar.org/CorpusID:252878844
9. Sample bioinformatics tool. https://bio.tools/bwa. Accessed 18 June 2025
10. Candela, L., Mangione, D., Pavone, G.: The fair assessment conundrum: reflections on tools and metrics. Data Sci. J. **23**, 33 (2024). https://api.semanticscholar.org/CorpusID:270073165
11. Clarke, D.J., et al.: FairShake: toolkit to evaluate the fairness of research digital resources. Cell Syst. **9**(5), 417–421 (2019)
12. Corcho, Ó., Ekaputra, F.J., Heibi, I., Jonquet, C., Micsik, A., Peroni, S., Storti, E.: A maturity model for catalogues of semantic artefacts. Sci. Data **11** (2023). https://api.semanticscholar.org/CorpusID:258615711
13. CSIRO MISC questionnaire. https://web.archive.org/web/20210813120307/http://5stardata.csiro.au/. Accessed 18 June 2025
14. Data Archiving and Networked Services (DANS): SATIFYD online questionnaire (2019). https://satifyd.dans.knaw.nl/. Accessed 22 Apr 2024
15. David, R., et al.: Fairness literacy: the Achilles' heel of applying fair principles. CODATA Data Sci. J. **19**(32), 1–11 (2020)
16. Devaraju, A., Huber, R.: An automated solution for measuring the progress toward fair research data. Patterns **2**(11), 100370 (2021). https://doi.org/10.1016/j.patter.2021.100370
17. F-UJI MISC tool. https://www.f-uji.net/index.php?action=test. Accessed 18 June 2025
18. Fair-checker tool. https://fair-checker.france-bioinformatique.fr. Accessed 18 June 2025
19. Foops! misc tool. https://foops.linkeddata.es/FAIR_validator.htm. Accessed 18 June 2025
20. Gaignard, A., Rosnet, T., De Lamotte, F., Lefort, V., Devignes, M.D.: Fair-checker: supporting digital resource findability and reuse with knowledge graphs and semantic web standards. J. Biomed. Semant. **14**(1), 1–14 (2023). https://doi.org/10.1186/s13326-023-00289-5
21. Garijo, D., Corcho, O., Poveda-Villalón, M.: FOOPS!: an ontology pitfall scanner for the fair principles. In: International Semantic Web Conference (ISWC) 2021. Posters, Demos, and Industry Tracks. CEUR Workshop Proceedings, vol. 2980. CEUR-WS.org (2021). http://ceur-ws.org/Vol-2980/paper321.pdf
22. GDPR checklist. https://gdpr.eu/checklist/. Accessed 19 June 2025
23. Peters-von Gehlen, K., Höck, H., Fast, A., Heydebreck, D., Lammert, A., Thiemann, H.: Recommendations for discipline-specific fairness evaluation derived from applying an ensemble of evaluation tools. Data Sci. J. **21**, 7–7 (2022)

24. Genova, F., et al.: Recommendations on FAIR metrics for EOSC. Publications Office of the European Union (2021)
25. Gene ontology. https://www.ebi.ac.uk/ols4/ontologies/go. Accessed 22 Apr 2025
26. Sample Gouvernemental dataset. https://www.data.gouv.fr/en/datasets/donnees-relatives-a-lepidemie-de-covid-19-en-france-vue-densemble/. Accessed 18 June 2025
27. Sample harvard dataset, https://dataverse.harvard.edu/dataset.xhtml?persistentId, https://doi.org/10.7910/DVN/JGO6VI. Accessed 22 Apr 2024
28. Huerta, E.A., et al.: Fair for AI: an interdisciplinary and international community building perspective. Sci. Data **10** (2022). https://api.semanticscholar.org/CorpusID:260201856
29. Jones, S., Grootveld, M.: How fair are your data? (2021). https://doi.org/10.5281/zenodo.5111307
30. Sample KAGGLE dataset. https://www.kaggle.com/datasets/imdevskp/coronavirus-report. Accessed 18 June 2025
31. Krans, N., Ammar, A., Nymark, P., Willighagen, E., Bakker, M., Quik, J.: Fair assessment tools: evaluating use and performance. NanoImpact **27**, 100402 (2022)
32. Lamarre, P., Andersen, J., Gaignard, A., Cazalens, S.: A generic framework to better understand and compare fairness measures. In: Alam, M., Rospocher, M., van Erp, M., Hollink, L., Gesese, G.A. (eds.) Knowledge Engineering and Knowledge Management - 24th International Conference, EKAW 2024, Amsterdam, The Netherlands, 26-28 November 2024, Proceedings. LNCS, vol. 15370, pp. 291–308. Springer (2024). https://doi.org/10.1007/978-3-031-77792-9_18
33. Li, B., et al.: Trustworthy AI: from principles to practices. ACM Comput. Surv. **55**, 1–46 (2021). https://api.semanticscholar.org/CorpusID:238259667
34. Linked open vocabularies (lov). https://lov.linkeddata.es/dataset/lov. Accessed 18 June 2024
35. Maillot, P., et al.: An open platform for quality measures in a linked data index. In: Chua, T., Ngo, C., Lee, R.K., Kumar, R., Lauw, H.W. (eds.) Companion Proceedings of the ACM on Web Conference 2024, WWW 2024, Singapore, Singapore, 13-17 May 2024, pp. 1087–1090. ACM (2024). https://doi.org/10.1145/3589335.3651443
36. Sample MOODLE course. https://moodle.polytechnique.fr/course/index.php?id=1018. Accessed 18 June 2025
37. Moser, M., Werheid, J., Hamann, T., Abdelrazeq, A., Schmitt, R.H.: Which fair are you? A detailed comparison of existing fair metrics in the context of research data management. In: Proceedings of the Conference on Research Data Infrastructure, vol. 1 (2023)
38. O'faire misc tool. https://agroportal.lirmm.fr/landscape#fairness_assessment. Accessed 18 June 2025
39. Sample PANGAEA dataset. http://doi.org/10.1594/PANGAEA.908011. Accessed 18 June 2025
40. Poveda-Villalón, M., Espinoza-Arias, P., Garijo, D., Corcho, Ó.: Coming to terms with fair ontologies. In: International Conference Knowledge Engineering and Knowledge Management (2020). https://api.semanticscholar.org/CorpusID:225078634
41. Sample rdf metadata. https://data.rivm.nl/meta/srv/eng/rdf.metadata.get?uuid=1c0fcd57-1102-4620-9cfa-441e93ea5604&approved=true. Accessed 18 June 2025

42. Slamkov, D., Stojanov, V., Koteska, B., Mishev, A.: A comparison of data fairness evaluation tools. In: Budimac, Z. (ed.) Proceedings of the Ninth Workshop on Software Quality Analysis, Monitoring, Improvement, and Applications, Novi Sad, Serbia, September 11-14, 2022. CEUR Workshop Proceedings, vol. 3237. CEUR-WS.org (2022). https://ceur-ws.org/Vol-3237/paper-sla.pdf
43. Sun, C., Emonet, V., Dumontier, M.: A comprehensive comparison of automated fairness evaluation tools. In: 13th International Conference on Semantic Web Applications and Tools for Health Care and Life Sciences, pp. 44–53 (2022)
44. Sample training material. https://tess.elixir-europe.org/materials/make-your-research-fairer-with-quarto-github-and-zenodo. Accessed 18 June 2025
45. Van De Sompel, H., Soiland-Reyes, S.: Fair signposting: exposing the topology of digital objects on the web. In: International FAIR Digital Objects Implementation Summit 2024. TIB Open Publishing (2024)
46. de Visser, C., et al.: Ten quick tips for building fair workflows. PLOS Comput. Biol. **19** (2023). https://api.semanticscholar.org/CorpusID:263224298
47. de Vries, A.: The growing energy footprint of artificial intelligence. Joule (2023). https://api.semanticscholar.org/CorpusID:264050478
48. Sample who dataset. https://data.who.int/dashboards/covid19/data. Accessed 18 2025
49. Wilkinson, M.D., et al.: The FAIR guiding principles for scientific data management and stewardship. Sci. Data **3**(1), 1–9 (2016)
50. Wilkinson, M.D., et al.: Evaluating fair maturity through a scalable, automated, community-governed framework. Sci. Data **6**(1), 174 (2019)
51. Wilkinson, M.D., Sansone, S.A., Marjan, G., Nordling, J., Dennis, R., Hecker, D.: FAIR assessment tools: towards an "apples to apples" comparisons (2023). https://doi.org/10.5281/zenodo.7463421

Jointly Trading Energy and Flexibility Based on FlexOffers and Blockchain

Laurynas Siksnys[1,2] and Torben Bach Pedersen[1,2(✉)]

[1] FlexShape Aps, 6000 Kolding, Denmark
laurynas@flexshape.dk, tbp@cs.aau.dk
[2] Aalborg University, Selma Lagerløfsvej 300, 9220 Aalborgø, Denmark

Abstract. Traditionally, electricity is traded in some markets, e.g., day-ahead spot, while flexibility is traded in other (regulating power, ancillary services) markets. With the advent of distributed electricity production, energy communities and flexible loads like heat pumps and Electric Vehicles (EVs), there is an unmet need to consider energy and flexibility as two sides of the same coin and trade them jointly instead of separately. This paper describes a solution for joint trading of (electrical) energy and flexibility, e.g., within energy communities. The solution is based on the so-called FlexOffer concept, which models energy needs and their associated flexibilities in one joint data object. The paper explains the actors and procedures for trading and market clearing. It also describes an implementation based on the IBM HyperLedger Fabric blockchain platform and experimental results from a pilot in the FEVER EU Horizon 2020 project.

Keywords: Energy flexibility · energy trading · P2P · blockchain · distributed ledger · FlexOffers

1 Introduction

Traditionally, (electrical) energy has been traded in some markets, like the day-ahead spot and intraday markets in Europe, while flexibility, the ability to regulate production or consumption up or down in short periods, has been traded in other markets like Frequency Containment Reserve (FCR) and Automatic Frequency Restoration Reserve (aFRR) for regulating power/ancillary services. Given the rapid deployment of distributed energy resources (DERs), e.g., renewable generation from photovoltaic (PV) panels and wind turbines, new flexible consumption from Electric Vehicles (EVs) and heat pumps, and new players like energy communities (ECs), there is a need to re-organize the markets to better support the bottom-up local nature of the emerging energy system with higher amounts of flexible or highly varying production and consumption. Such emerging energy systems with increased energy variability and flexibility open up new opportunities at lower ends of the power grid, e.g., avoiding local grid congestion or offering lower electricity prices in behind-the-meter setups like energy communities.

This paper describes the development of a market platform for jointly trading energy and flexibility, developed in the FEVER EU Horizon 2020 project. The marketplace offered by the platform is based on the concept of FlexOffers (FOs) [1], a general and scalable model for energy flexibility which models the energy needs of a, potentially aggregated, load along with its inherent flexibilities in both time and energy amount. The market is designed to handle several user stories (USs) for an energy community, its members (peers), and the interaction with the surrounding energy grid based on Peer-to-Peer (P2P) principles. The relevant user stories are: *US1: Peer buying locally produced electricity from other EC peers; US2: Peer selling PV energy to other EC peers; US4: Trading directly between a Distribution System Operator (DSO) and EC peers; and US5: Multi-EC trading between a DSO, EC-Operator, and EC peers.* Table 1 below summarizes the user stories US1–5 by focusing on "what", "who", and "why" aspects.

Table 1. User Story Details

#	What is traded?	Who is trading?	Why is it traded?
US1	Amount of electricity for a specific appliance	EC member/buyer peer	Use locally produced electricity (save money, reduce CO_2, support EC)
		EC member/seller peer	Sell excess electricity (earn/save money, support EC)
US2	Amounts of excess PV-generated electricity	EC member/buyer peer	Use locally produced electricity (save money, reduce CO_2, support EC)
		EC member/seller peer	Sell excess electricity (earn/save money, support EC)
US3	Flexibility (up/down variations of energy/power)	EC peers	See US4 and US5 below
US4	Flexibility (up/down variations of energy/power)	DSO	Stabilize the EC grid
		EC members	Reduce EC grid maintenance costs
US5	Flexibility (up/down variations of energy/power)	DSO	Stabilize the DSO grid
		EC-Operator	Bring value to EC (from DSO)
		EC members	Reduce costs

The paper makes the following contributions: 1) Specifying how FOs can be used both as *energy* and *flexibility* products and thus form the foundation of joint trading of energy and flexibility; 2) Proposing market concepts, actors and mechanisms such as market clearing for such a joint market based on P2P principles; 3) Presenting an implementation of the marketplace supporting functionality in the *FlexTrading DAPP* (Distributed Application) based on the IBM Hyperledger Fabric blockchain (distributed ledger) framework; 4) Summarizing results from the pilot in the FEVER project.

The rest of the paper is organized as follows. Section 2 describes how FOs are used as joint energy/flexibility products. Section 3 describes market concepts, actors, and mechanisms. Section 4 describes the pilot implementation and experiments, while Sect. 5 concludes and points to future research.

2 Related Work

2.1 Flexibility and FlexOffers

FlexOffers (then called Flexibility Objects) were first introduced as a computer science concept in a conference paper [17] and its journal extension [22]. Here, the focus was on methods for aggregation and disaggregation. Methods for using aggregated flexibility for balancing were proposed in [18]. Methods for aggregation that respect grid constraints were proposed in [23, 24].

Methods for prediction of flexibility based on device-level data were first proposed in [19]. Later, methods for generating and optimizing FlexOffers based on predicted flexibility were proposed for white appliances and EVs [25], heat pumps [25, 31, 34, 39], and batteries [30]. Methods for converting flexibility from one energy vector (electricity) to another (heat) and back were proposed in [33].

Methods for adaptive user-oriented direct load control of flexible residential devices were proposed in [28].

Trading of flexibility using FlexOffers was considered in several papers. Specifically, the value of flexibility for trading in regulation markets was quantified in [20]. Methods for trading aggregated flexibility as so-called flexible orders in the day-ahead spot market were proposed in [26]. Methods for utilizing device-level demand forecasting for trading in flexibility markets were proposed in [27]. Flexibility trading in bottom-up cellular energy systems was considered in [29].

Metrics for measuring the amount of flexibility were first proposed in [21] and recently significantly extended in [37].

The key aspect of the uncertainty about the amount of flexibility actually available at delivery time was considered in [32, 35].

To support easier integration of FlexOffers in other systems, a recent paper [37] significantly extended the popular SAREF4ENERGY ontology to provide full support for even advanced FlexOffers.

2.2 Traditional Energy and Flexibility Trading

Other related work can be grouped according to two dimensions: a) trading energy or flexibility and b) using blockchains or not. Traditional energy markets such as Nordpool Spot [5] or EPEX Spot [6] trade only energy, not flexibility, and are not based on blockchains. Traditional TSO-level flexibility markets such as aFRR [15] or FCR [16], as well as more recent DSO-level local flexibility markets [7], for example NODES Market [8] and the GOFLEX market [13], trade only flexibility, not energy, and are not based on blockchains.

2.3 Blockchain-Based Energy and Flexibility Trading

The blockchain-based markets from Powerledger [11] and Grid Singularity [12] trade energy only while the market in [40] trades energy *data* on blockchains. The blockchain-based markets from Electron [9] and the PLATONE EU project [10] trade flexibility only. The paper [41] considers securing energy data for trading from Denial-of-Service attacks.

In summary, we believe this paper to be the *first* to support *joint* trading of both energy and flexibility based on blockchains. A brief summary of a few of the high-level ideas was first presented in a 2-page poster paper [38].

3 FlexOffers as Energy and Flexibility Products

3.1 FlexOffer and Product Overview

FlexOffers are adopted in FEVER as a common unified representation of energy flexibilities, enabling FEVER system interoperability and the reuse of flexibility management algorithms, libraries, and tools. FlexOffers are used to define both selling and buying bids of the P2P marketplace (FlexTrading DAPP). We now discuss the two types of products which can be specified by an FO: ***PE: Energy Product*** - a series of electricity amounts (in kWh) consumed/produced within consecutive time intervals (e.g., of 15 min duration). The full series can, potentially, be shifted in time within the user-given bounds (during matching) – to improve the buyer's or seller's position (utility) – e.g., US1–2 above. ***PF: Energy Flexibility Product*** - a series of electricity amount deviations (deltas) that represent increased or reduced production or consumption (of a prosumer, in ΔkWh) within consecutive time intervals (e.g., of 15 min duration). The full series can, potentially, be shifted in time within the user-given bounds (during matching) – to improve the buyer's or seller's position (utility). For simplicity, these product instances below in the text are called PE and PF products.

A single peer can offer and/or request one of these products by submitting a bid, i.e., a FlexOffer. Normally, this should be done in an automatic fashion through an agent system. In both cases, ***FlexCoins*** (*an EC-internal crypto-pseudo-currency with an exchange rate of 1 euro also developed in FEVER*[2]*)* are used to define the value of PE and PF. The products PE and PF are separated and traded in two different phases (see Appendix A.3). FlexTrading acts as a multi-product marketplace. Product PE-PF FOs are indivisible (i.e., are "take it or leave it"). This means that energy amounts and prices of all FO slices are decided by FlexTrading if a bid is accepted. This property means that a peer will be able to guarantee/plan power delivery ahead in time for the full duration of a process in question (e.g., EV charging, washing dishes, etc.). A single FO defines PE or PF using time, energy amount, location, and price parameters and capture complex inter-dependencies between them, e.g., by utilizing advanced forms of constraints supported by an FO [3]. We now discuss how exactly these products are represented by FlexOffers.

3.2 FlexOffers as Energy Products

Figure 1 exemplifies a single FO, representing *PE*. In this example, a peer expresses electricity demand and supply amounts (in kWh) for 3 consecutive time intervals (of 15min), denoted as slices. The starting time of energy delivery (and consumption) is not fixed and can be varied within the permitted FO time flexibility interval, specifically from 09:00 (earliest start time) to 10:00 (latest start time) with 10:45 as the latest end time (see black dashed lines). The offered/requested energy amounts of these 3 slices are also not fixed and can be varied within the respective energy flexibility intervals (yellow areas). The first 2 consecutive slices additionally define minimum energy amounts to be consumed (1st) and generated (2nd), respectively (orange areas). Each such interval also defines a price curve, which specifies the concrete amount of energy a peer wishes to buy or sell as a function of energy price (in FlexCoins) he/she will be offered for this interval. When submitted, FlexTrading will match this bid-FO with bid-FOs from other peers, by exploring a variety of allocation (start time) options. Ultimately, a schedule will be generated, which specifies concrete demand and supply start time (red arrow), energy amounts (v1-v3), and selling/buying price (in FlexCoin) for each time interval.

For PE, no seller-side and buyer-side bids are distinguished. Based on a bid submitted, FlexTrading will automatically decide whether a particular peer acts as a buyer or a seller (of electricity) at a particular time interval. Unlike bids in the traditional auction-based electricity markets, a FO may capture complex inter-dependencies between slices (time units) [3], which makes PE fundamentally different compared to the majority of solutions discussed in the literature. When bid FOs have only 1 slice each, trading in the PE marketplace provided by FlexTrading is similar to a double auction with sealed bids.

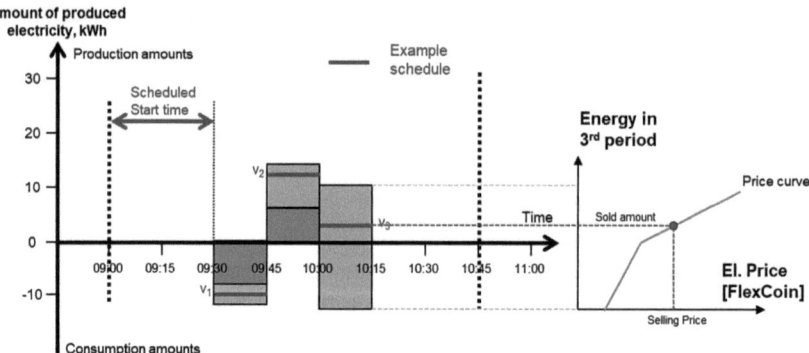

Fig. 1. Example FO with price info (for 3rd slice) for PE. Reproduced from [38].

3.3 FlexOffers as Flexibility Products

The *product PF* is modelled in a similar way, but with a few significant differences and additions in FO parameters. Specifically, FOs of the PF product define so-called baseline schedules and reward attributes (details shown in Fig. 5 in Sect. 5.1). The baseline schedule (green lines) indicates a socio-economically optimal electricity consumption/generation schedule of a peer, whereas reward attributes indicate possible deviations from this baseline depending on a reward offered to the peer for such a flexibility service. For example, energy shift reward for the FO slice 3 in Fig. 2 indicates the peer's possibility to further increase or reduce production in the range from 7 to 30 kWh depending on the reward offered, with the baseline value being 15 kWh. Likewise, a peer may indicate time shift reward that enables several time-shifting options within the full FO time shifting period (see yellow lines at the bottom) depending on the time shifting reward offered. Other types of rewards are possible, e.g., commitment reward which indicates the peer's requested amount of FlexCoins for fulfilling a prescribed schedule (irrespective to whether it represents a baseline or a deviation). Note, without such a commitment reward otherwise, a peer is allowed to consume/produce electricity in any desirable way. For PF, buyers and sellers are distinguished. This means that the sender of a FO (bid) needs to indicate whether he/she requests or offers such a flexibility service.

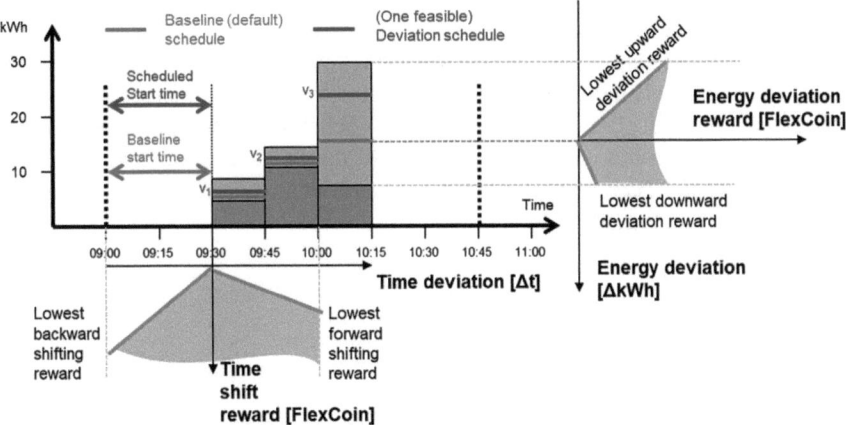

Fig. 2. Example FO with price info (for 10:00–10:15) for PF. Reproduced from [38].

3.4 Energy Community Peer Bidding Strategy.

The FO representation of bids give a wide range of options for peers to specify PE and PF products. FOs can be generated (selectively) for a single or multiple consecutive time intervals (slices), with or without minimum (/maximum) requested total and/or interval energy, simple or advanced price curves and/or reward parameters. A specific approach of a FO generation is specified manually by the user as a bidding strategy (settings).

4 The P2P Marketplace

4.1 Market Organization

The market for products PE and PF is organized as a number of distinct independent groups (bidding areas), each representing a physical EC. The groups are considered flat (i.e., following the "copperplate" model) [4], which means that neither grid capacity limits nor grid fees and costs of transferring energy between group peers are considered.

A peer may trade in one or more such groups. For trading in a specific group, a peer (user) has to hold a digital contract between the peer and an EC (group) stored on the blockchain. Such a contract specifies terms and conditions applicable in this specific group - set by the EC operator (e.g., allowed imbalances and their fees). Each traded PE and PF FO should include a (spatial) parameter indicating the concrete market group (bidding area) the product is traded for. Up to three *business actor roles* are involved in every group: *EC Operator*, *DSO Peer*, and *EC Peer*. An EC Operator is an internal entity with no business interest in trading. It helps the EC as a whole by facilitating trading between DSO and EC peers and manages independent auctions for PE and/or PF products. A DSO Peer is a local DSO which may buy and activate PF products, each with a geo-tag and group (bidding area) identifier, to manage congestions within these groups. An EC Peer is a party connected to the EC grid, consuming and producing electricity with a valid FlexCoin account, seeking both private (e.g., save money) and as communal objectives (e. g., maximize the EC's self-consumption). An EC Peer is equipped with smart or sub-meters and is willing to share their (near) real-time measurements with EC Operator. In addition, it is equipped with an automated flexibility service providing/consuming (trading) agent (FSxA), which controls indirectly or directly (through an energy management system) electrical loads and can place bids on the EC peer's behalf based on e.g., bidding strategy and prices. Based on these assumptions, we split the EC Peer actor into two logical actors – *Human EC Peer* and *EC Peer FSxA*, and identify key information objects, associated to each in Fig. 3 below.

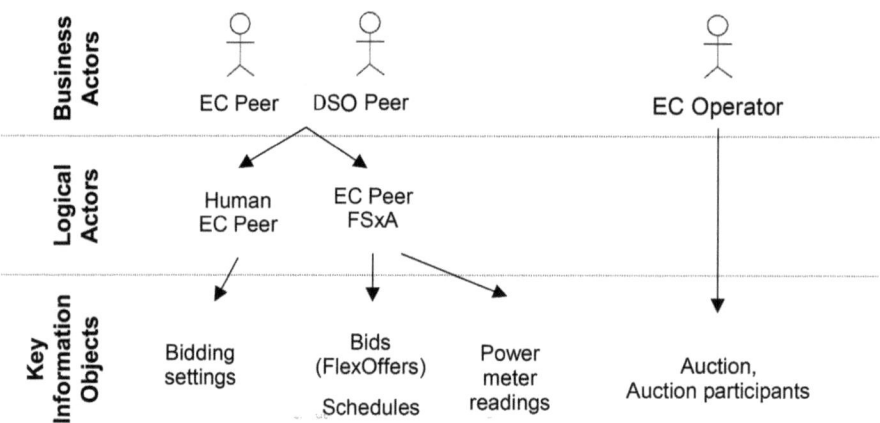

Fig. 3. P2P marketplace actors and information objects

4.2 Market Clearing Mechanism

FlexTrading continuously receives PE-PF FOs and groups them based on a product type and a targeted EC. Finally, within each group, the system matches them by repeatedly solving a complex optimization problem, which tries to find the best techno-economical solution (see Pricing below) within some time horizon of T. When solving this problem, total energy and cost balance constraints, as well as price curves (and reward attributes) and inherent FO constraints of all participating peers are considered. To avoid infeasibilities when solving this optimization problem, the EC Operator includes so-called default bids (FOs), which represent amounts of energy imported from the grid (/exported to the grid) to EC (less favorable). Clearing/contracting, however, is performed when no further (re-)scheduling of a FO is allowed – this is explicitly indicated by the FO's latest assignment time parameter. Contracted energy amounts are typically defined for fixed quarter-aligned time intervals (e.g., 1:15 to 1:30) - thus matching FO slice specifications. This overall process is visualized in Fig. 4.

As seen in Fig. 4, a received FO (bid) goes through 3 states: **Submitted** bids are those which were checked, accepted, and scheduled (potentially several times). Bids remain in this submitted state until the (inside the FO) indicated *latest assignment time* is due. This FO parameter is peer (user)-specified and often matches the value of *earliest start time* for quickly dispatchable loads. **Contracted** bids can no longer be (re-)scheduled. The last computed FO schedule is considered final and is written into the distributed ledger. Thus, it becomes the subject of the peer's (closed) contract and the peer is required to deliver and consume energy amounts according to this contracted schedule. **Settled** bids are in their final state of the FO lifecycle. Required funds (of FlexCoins) are transferred to/from the peer's *personal account* into a *market pool account* (due to *peer-to-community* contracting) managed by the *EC operator*. These transactions are fully automated (since they are governed by so-called smart contracts). A bid enters the *settled* state either if measurements from registered smart meters (main or sub-meters) are received for the contracted FO period, or no measurements from a peer (meter) are available by a predefined deadline (which renders a delivery failure). In its nature, our proposed market concept is a variant of continuous auction model, where a marketplace tries to clear the market each time a new order is placed, instead of clearing at discrete time instances.

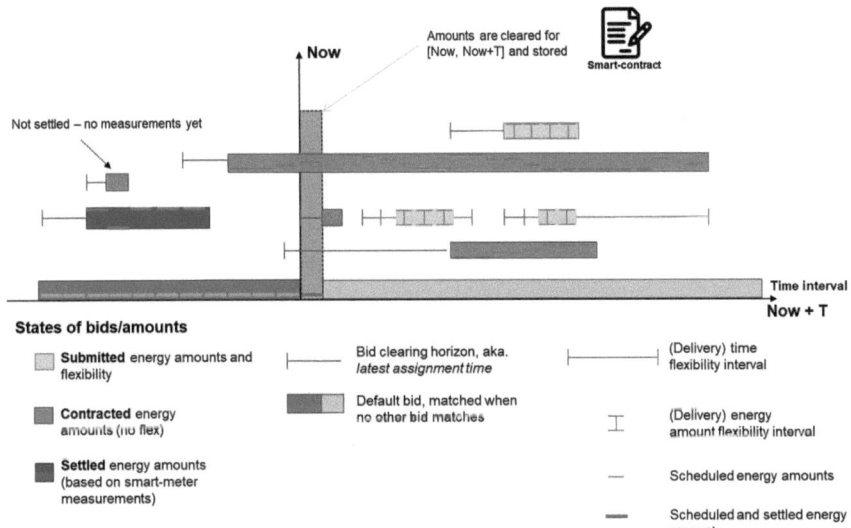

Fig. 4. FlexTrading market clearing mechanism

4.3 Pricing

For each market group (bidding area), *FlexTrading DAPP* allows EC operators to configure the following pricing-related parameters: **Pricing mechanism** - defines how prices of P_E products are established: *Pay-as-bid* - a peer gets its offered price (for a specific time interval); *Pay-as-cleared* (aka "uniform pricing") - all peers receive the same price (market-clearing price), even if they have offered different prices. **Energy source ranking** - defines how available energy sources (and sinks) are ranked during the matching process. By default, *merit order* (sorting by price in FlexCoins) is used. However, sorting based on energy sources are possible, i.e., that some sources of electricity (e.g., from RES) are prioritised and some are penalized (e.g., fossil-based) inside a EC.

4.4 Settlement

After a FO is executed and energy is consumed and produced for the complete duration of a FO, measurements from registered and EC-Operator smart meters (or peers) are collected. Based on these measurements, peer bids are settled and funds (in FlexCoin) are automatically transferred between participants. In the case of imbalances (differences between scheduled and measured amounts), FlexTrading will automatically provision surplus or deficit energy from/to a back-up energy source (e.g., the grid, less favourable). Also, pre-defined penalties (in FlexCoin, specified in the contract) are automatically applied to the peer, while also reducing the peer's performance index. This performance index can, potentially, be considered during energy source ranking (see Pricing above) and can increase or reduce the peer's trading position (for getting better prices for PE and PF products). Finally, EC-Operators will be provided a view of measurements from different EC peer meters within trading periods so electricity amounts traded in the P2P marketplace are excluded (not accounted for) in the final electricity supplier's bill.

Table 2. Essential methods (transactions) of the FlexTrading DAPP smart contract

Method Name	Method description
createAuction()	Creates a new instance of an auction
createParticipant()	Grants a new participant access to an auction instance
createMeter()	Registers a new power meter of an EC peer
placeBid()	Places a bid onto a specific instance of an auction. The method immediately returns a preliminary schedule, which includes prescribed energy amounts and preliminary prices
updateAuction()	Reschedules all submitted (not yet contracted) bids. Clears the auction by bringing all pending FlexOffers (FOs, i.e., those which latest assignment time is due) into the contracted state
reportMeasurements()	Upload EC peer's power meter measurements on chain. This is followed by the automatic validation of peer's contracted bids and settlement, during which specific amounts of FlexCoins is transferred between the EC-Peer's and the EC-Operator's FlexCoin accounts

5 P2P Marketplace Implementation Details

5.1 Data Model

We will start with defining core information objects for use by the FlexTrading DAPP. As seen in Fig. 5, FlexTrading builds up on a number of information objects from others DAPPs (Community Management DAPP and FlexCoin DAPP) and introduces some new ones. The most essential of those are discussed next.

Auction – this class specifies an instance of a marketplace. It includes a unique Id of the marketplace (*actionId*), an Id of energy community it is linked with (*ECid*), a type of product (P_E or P_F) to be traded (*productType*), pricing and ranking mechanisms to be used (*pricingMechanism, ranking*).

AuctionParticipant - this class specifies a participant, which has (was granted by the EC Operator) access to a specific marketplace instance. The participant is linked with a particular FlexTrading DAPP user (via *userId*) and its FlexCoint account (via *coinAccountId*). Specific parameters of the participant, which were agreed with the EC operator (e.g., *allowedImbalance*), are also represented by this class.

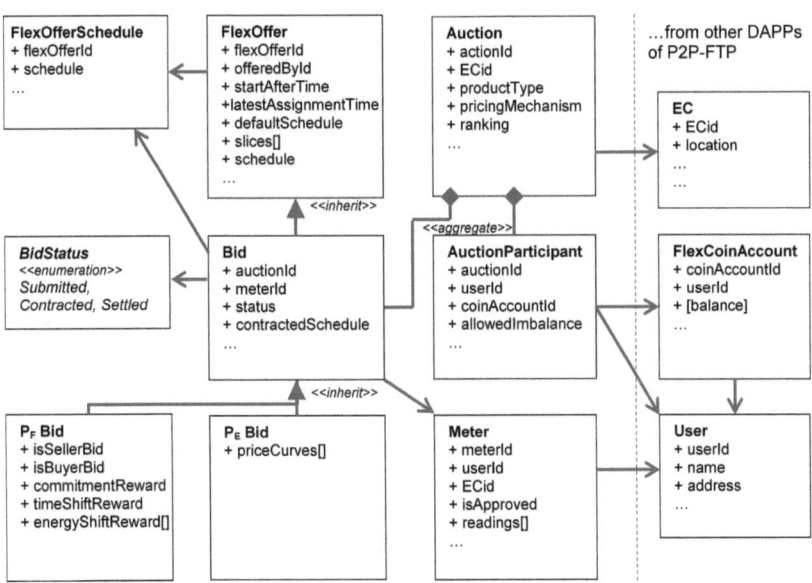

Fig. 5. Simplified data model of the FlexTrading DAPP

FlexOffer – this class specifies the EC peer's energy need/offer and associated flexibility, while covering a number of consecutive time intervals. The flexibility is defined by a number of constraints, including those defined by FlexOffer slices.

FlexOfferSchedule – this class specifies concrete energy amounts to be consumed and/or produced by the EC peer. The schedule is always linked with an instance of a FO (*flexOfferId*). Concrete energy amounts (*schedule*) are defined for each slice of the FO and they respect all FO constraints.

Meter – this class specifies an energy meter of an EC peer. It can be a certified smart meter, or a sub-meter. In the latter case, it needs to be manually approved by the EC operation to be able to use it for P2P trading.

Bid – this class extends the original FlexOffer class and specifies a bid targeting a particular market instance in the generic form. The bid and has several additional attributes, not available in the FlexOffer class. For example, *auctionId* links a bid with a particular marketplace instance; *status* specified the state of the bid (submitted, contracted, settled); *contractedSchedule* specifies concrete energy amounts to be consumed/produced by the participant. Two specializations of this class are possible: P_E Bid and P_F Bid.

P_EBid – this class specifies an offer or a request of the P_E product. It further extends the Bid with price curves (priceCurves) for each slice of a FlexOffer.

P_FBid – this class specifies an offer or a request of the P_F product. It further extends the Bid with a number of attributes relevant to P_F: whether the bid expresses an offer or a request of P_F, i.e., a flexibility service (*isSellerBid, isBuyerBid*); reward attributes of the offered/requested (*commitmentReward, timeShiftReward, energyShiftReward*).

5.2 Interaction Between Core Components

The smart contract of FlexTrading supports a number of methods (transactions) which will manipulate these information objects on-chain – see the Table 2 below.

Before P2P trading can take place, the EC-Operator needs to create one (or more) auctions (*createAuction*), register its participants (createParticipant) as well as their power meters (createMeter). For (an) already created auction(-s), the EC-Operator is able to edit (or delete) auctions, participants, meters (methods for that are not shown in the table). After the configuration is complete, P2P trading can take place in an automatic fashion. A simplified interaction between the FSxA of EC-Peer and the FlexTrading and FlexCoin DAPP smart contracts during this automatic process is shown in Fig. 6 below.

Initially, the FSxA constructs a bid with all required FlexOffer parameters based on the bidding settings set my Human EC-Peer. The bid is then placed on the blockchain and handed by the FlexTrading DAPP smart contract. The smart contract automatically (and deterministically) matches this bid with the other bids within the auction and returns a preliminary schedule with indicative start time, energy amounts, and prices. Periodically, all auction bids are rescheduled. As a result, some bids receive preliminary schedules, the rest – contracted schedules with the final start time, energy amounts and prices. Finally, FSxA periodically reports power measurements to validate the execution of the contracted bids. Successfully validated bids, or bids without measurements available (in due time), trigger an automatic transfer of FlexCoin amounts from the EC-Peer to EC-Operator or vice versa.

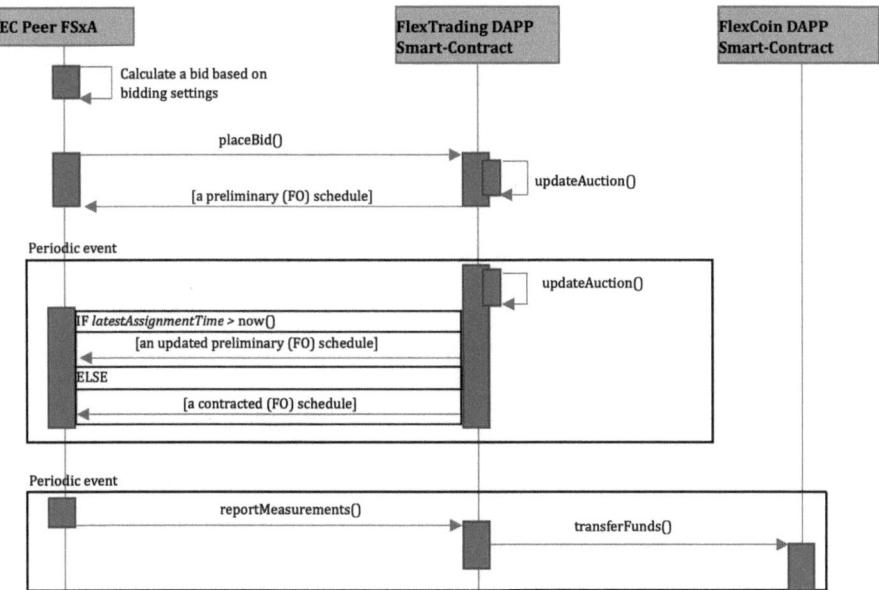

Fig. 6. Interaction between FSxA and smart contract

5.3 Topology of Hyperledger Fabric Network

The smart contract presented in the earlier section (together with other relevant components) will be deployed on the HLF network. An example minimalistic deployment topology is shown in Fig. 7 below.

Initially, all actors taking part in P2P trading within a single EC form a *consortium* – which may include EC peers, EC Operator, and some 3rd party organization like a local DSO or a software service provider. For P2P trading in a specific EC, the members of the consortium are connected to the same common HLF *channel* ("channel" in the figure), which acts as a main communication medium allowing them to communicate to each other. The channel should be configured in such a way that only the EC-Operator can add new members to it. However, a less centralized channel configuration where a pre-defined share of all members has to agree in order to add new market participants is also possible.

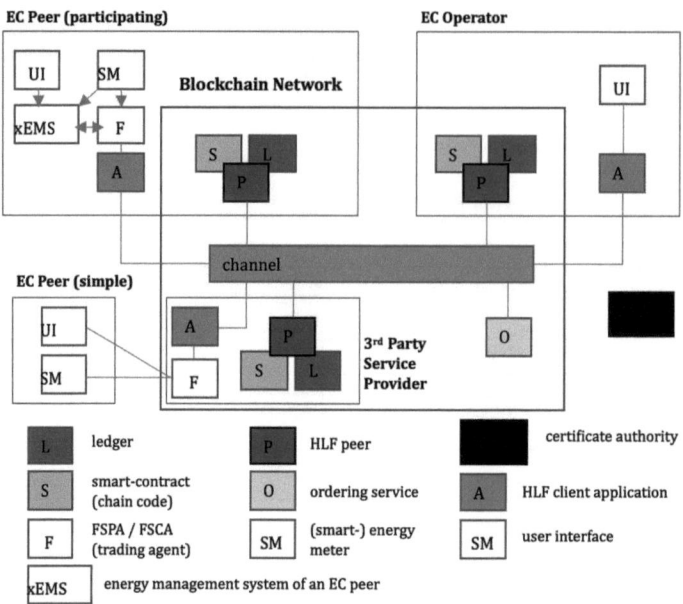

Fig. 7. Topology of HLF network of FlexTrading

As seen in the figure, the network includes a number of HLF network (*endorsing*) peers ("P" nodes), managed by different consortium members. For example, the EC Operator runs a HLF peer, which manages a copy of the FlexTrading DAPP's smart contract ("S" node) and the ledger ("L" node). For resilience and availability, the EC Operator may run more HLF peers in other deployment configurations. In addition, the EC Operator runs a HLF *client application ("A" node)*, which (1) offers a user-interface ("UI" node) to a (human) user for managing the marketplace and is able to (2) send transaction proposals onto the channel by taking advantage of a Software Development Kit (SDK) for the Hyperledger Fabric. Similarly, two kinds of EC Peers are envisioned:

EC Peer (participating) - pro-actively participates in the HLF network. It runs a *HLF peer ("L" node)*, which manages a copy of the FlexTrading DAPP smart contract and the ledger ("S" and "L" nodes). In addition, the participating peer has an *energy management system* ("xEMS" node) with an integrated (or an external) FSxA ("F" node) which acts as a *trading agent* towards the smart contract with ability to send transaction proposals via a HLF client application ("A" node). The smart-meter ("SM" node) measurements are accessed by both the energy management system as well as the FSxA for load monitoring and bid validation purposes (ST1.4-ST1.7). In real-world setups, the HLF client application, FSxA, and xEMS ("A", "F", and "xEMS" nodes) can be integrated into a single software system offered to the EC Peer.

EC Peer (simple) – participates in the HLF network through a selected 3rd party service provider. The 3rd party service provider hosts its own instances of FSxA and the HLF client application, as well as the HLF peer, and offers the EC peer an access to the HLF network though its own 3rd-party software platform, which integrates all these components. In this setup, EC peer can use a user interface ("UI" node) offered by the

3rd party platform but needs to provide an access to (the data of) a power meter (and a remotely controlled relay/device, e.g., a smart plug).

These exists a Certificate authority ("CA" node). This is an institution that issue *certificates* to consortium members and network nodes. These certificates define *user identities* and are used to map these identities to organizations, e.g., EC Operator. It is reasonable to assume that there exists a single common *Certificate Authority* that issues certificates to all – EC Operator, EC Peers, and 3rd party service providers. A blockchain network-wide external Certificate Authority might be additionally used, if needed.

Finally, the *ordering service* ("O" node) is responsible for ordering and packaging endorsed transactions into blocks - thus providing delivery guarantees. In production, this service should be provided by a cluster of nodes.

In order to transact, client applications ("A" nodes) submit transaction proposals to HLF (endorsing) peers and need to receive a predefined number of positive responses (i.e., endorsements) before they can send transactions to the ordering service. This number is fixed (in the endorsement policy) and embedded in the channel policy - defined at the moment of chain code instantiation. In order to maintain an efficient and scalable system, we recommend a policy where the signatures are required from: (1) a member peer associated with the EC Operator and (2) two or three other peers, belonging to different organizations.

5.4 Private Bids on Ledgers

In this section, we discuss a distribution of actual data stored in the ledgers of different peers. As discussed earlier, each HLF peer manages a *ledger* which stores all *transactions* from a particular blockchain *channel*. These interlinked transactions determine the *world state*, which is stored a *classical database* for increased speed and easy queries. However, in our P2P market design, a bid (FlexOffer) submitted by some EC peer should remain *sealed* from the rest and shared only with the EC Operator. This can be achieved in Hyperledger Fabric using so-called *private transactions,* which were introduced in Hyperledger Fabric version 1.2. With this feature, data considered private can be shared with only authorized organizations (e.g., of EC Operator). Most importantly, this feature keeps private data confidential from an ordering service, which may be controlled by an organization unauthorized to see the data. In FlexTrading DAPP, a bid (FlexOffer) is split into two transactions: (1) a *private transaction*, holding its full information (attributes); and (2) a *public transaction* containing only basic attributes (which do not have to be protected) and visible for all market participants. The full details of the bid are stored in a private state database and shared between the invoking EC peer and the EC Operator. A hash of the transaction is stored on the public blockchain for auditing purposes. This means that, in addition to the blockchain and the database storing the world state, every ledger of the participating peers additionally stores the (private) details of bids (Flex-Offers) in a private state database. The EC Operator peer(-s) thus maintains a number of such private state databases, each for a different participating HLF peer in the HLF network. Figure 8 exemplifies the content of the ledgers of EC peer 1, EC peer 2, and the EC Operator.

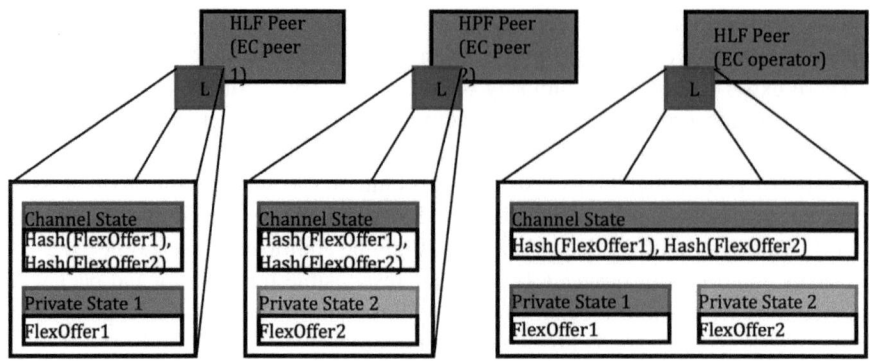

Fig. 8. Ledger (L) contents of HLF peer nodes that store private data

5.5 Nested P2P Trading

The presented P2P marketplace design is primarily suited for P_E and P_F trading within a single EC. In this section, we discuss how this P2P marketplace can be used to cater for the more advanced multi-EC trading use-case US5. Figure 9 shows a simplified model of such *nested trading*. Here, EC1 is an example local energy community, where peers trade P_E and P_F products with each other. Actor A_1 takes the role of EC1 operator and EC2 peer at the same time. Therefore, A_1 on the behalf of EC1 has a capability to offer/request energy (P_E) or flexibility service (P_F) products to/from the outer EC2. To trade on EC2, A_1 continuously submits P_E bids for energy and/or P_F bids for flexibility services onto the EC2 market (bidding area). On successful trades on the EC2 market, A_1 includes counter P_E and/or P_F bids into the EC1 market (bidding area).

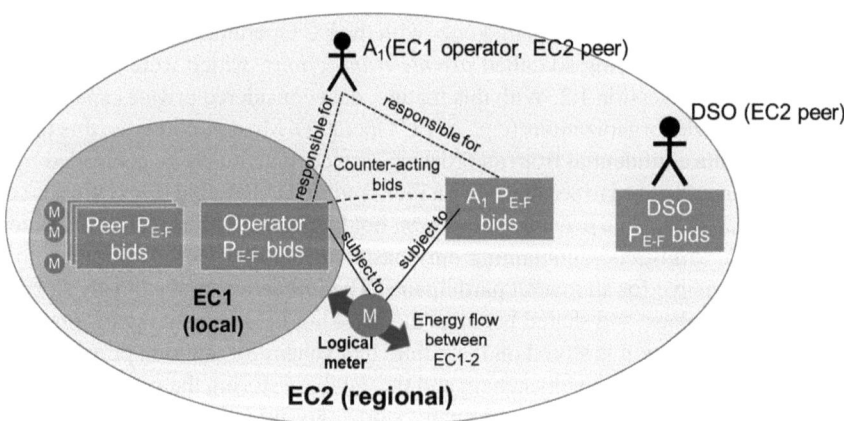

Fig. 9. Simplified model of energy flows between two nested ECes

6 Marketplace Clearing Details

6.1 Roles

The FlexTrading DAPP supports multiple concurrent *auctions*, where an *auction participant* takes one of the following two positions (roles) within an auction:

- **Flexibility Provider** - these are electricity consumers and/or producers (i.e., EC-Peers) who consume and/or produce pre-defined packets of electricity and are, potentially, flexible in terms of (1) when this consumption and/or generation should start; and (2) how much energy should be consumed and/or generated during consecutive (e.g., 15 min) intervals. Among these flexibility options, a flexibility provider has its own preferred electricity consumption/ generation profile, a so-called *default schedule* (baseline). The flexibility provider may request some *reward in FlexCoins* from a *flexibility consumer* (see below) for actively shifting its load in time and amount with respect to this default schedule (baseline). This should be done by respecting all predefined flexibility bounds. To define such an energy packet, flexibility bounds, the default schedule, and requested reward, so-called FlexOffers for P_F products are used.
- **Flexibility Consumer-** these are EC-Operators or specialized EC-Peers (including EC grid operators / DSOs) who are responsible for energy flows within an EC area and are willing to pay flexibility providers to have a right to shift their loads in time and amount, e.g., to avoid congestion or balance demand and supply within an EC. Such flexibility consumers compete in an eBay-like *auction* (formally *open ascending price auction, aka.* The *English auction*), by offering increasingly higher amounts of rewards to flexibility providers for the right to shift (schedule) their loads in time and amount.

Flexibility providers continuously submit their *flexibility provider bids* to an auction. These go through two consecutive auction phases: *flexibility dispatch phase* and *energy dispatch phase*.

6.2 Flexibility Dispatch Phase

In this phase, all flexibility provider bids are made open to flexibility consumers. Consequently, flexibility consumers can retrieve these flexibility provider bids, evaluate them, and propose a counter-offer by submitting a *flexibility consumer bid*. This flexibility consumer bid *(ask)* includes a schedule for a desired load execution (respecting all flexibility provider's constraints) and an offered reward in FlexCoins. If this reward is found to be the highest, the schedule of this highest bidder becomes a so-called *winning schedule* and is assigned for a physical dispatch (execution) by a specific flexibility provider.

Winning schedules or (non-winning) default schedules define specific energy amounts (in kWh) to be consumed and/or produced by flexibility providers within specific consecutive time intervals (e.g., of 15min duration). Therefore, they tell "how much" and "when" electricity should be consumed and/or generated. However, they do not tell "where this electricity comes from/goes to" and "what the price in FlexCoins is for each consumed/produced kWh of electricity". For this reason, all winning or (non-winning) default schedules of all flexibility providers are automatically transferred from the flexibility dispatch phase into the *energy dispatch phase*, described next.

6.3 Energy Dispatch Phase

In this phase, flexibility providers (EC-Peers) buy or sell electricity from/to each other close to real time at fixed-duration time intervals (e.g., of 15 min) according to the schedules from the flexibility dispatch phase. For this, flexibility provider bids are partitioned into fixed-duration *slices* in order to match specific clearing periods (e.g., of 15 min). For a specific clearing period (bid slice), flexibility providers compete in a double-sided energy dispatch (sub-)auction, similar to that of the traditional energy exchanges. The competition is based on sealed (closed) electricity prices, proposed by the flexibility providers in their respective flexibility provider bids - see FlexOffers for P_E products. For this, the merit-order mechanism is used: highest electricity consumer bid slices are matched with the lowest electricity producer bids (offers). Eventually, the auction is cleared at the equilibrium energy amount and price, aka. *Market clearing amount* and *market clearing price (MCP)*.

The auction owner (e.g., EC-Operator) can choose (at an auction creation time) one of the pricing mechanisms, which determine how final electricity prices are nominated to market participants:

- **Pay-as-clear** - participants are automatically awarded the price of the most expensive buyer bid accepted (matched)
- **Pay-as-bid** - participants are awarded prices from their individual bids.

Accepted electricity consumer and producer bid slices are *contracted* and, sometime later (e.g., after actual meter measurements become available or a timeout) *settled* (see Settlement below).

The auction owner can also decide how the remaining (unaccepted/unmatched) electricity consumer and producer bid slices are handled:

- **Contracted at predefined prices** All unmatched bid slices are satisfied at pre-predefined "highest buying price" and "lowest selling price".
- **Excluded for contracts** Bid slices are not contracted, and, therefore, they do not participate in the settlement process (see Settlement below).

In both cases, physically consumed and produced electricity of the remaining (non-winning) bid slices at those trading intervals will have to be handled outside the FlexTrading DAPP in the conventional way, i.e., through a traditional open *electricity contract* involving an electricity supplier (or EC-Operator acting on the supplier behalf).

In general, the auction owner (e.g., an EC-Operator) can enable or disable flexibility and/or energy dispatch phases (sub-auctions) and thus configure an auction instance to act as a *flexibility marketplace*, *energy marketplace*, or a (coupled) *flexibility and energy marketplace*. Thus, an auction instance can be tailored to the specific EC configuration and/or end-user application in mind.

6.4 Settlement Process

After the contracted electricity amounts are consumed and produced, the settlement process takes place.

The FlexTrading DAPP marketplace continuously receives measurements from pre-approved participant energy meters (smart meters, smart plugs, etc.). These measurements are used to automatically *settle* flexibility provider bids, and thus transferring respectively amounts of FlexCoins to/from FlexCoin accounts of the involved participants. Settlement is also a two-phase process. First, the bid is settled from the flexibility dispatch point of view, then, from the energy dispatch point of view.

Flexibility Settlement. If measurements show that the winning schedule was successfully followed (within a predefined error/imbalance in kWh), the respective flexibility provider bid is marked as *successfully executed* from the flexibility dispatch point of view. Consequently, respective amounts of FlexCoins are automatically transferred from the flexibility consumer's FlexCoin account to the flexibility provider's FlexCoin account.

Energy Settlement. Later, flexibility provider bid slices are settled from the energy dispatch points of view. Specifically, electricity consumers buy, at the contracted price (in FlexCoins), energy amounts they've *physically consumed* - but no less than what was contracted. Likewise, electricity producers sell, at the contracted price (in FlexCoins), energy amounts that they have *physically generated* - but no more than it was contracted. In both cases, FlexCoins are automatically transferred to/from the FlexCoin account of an auction participant (EC-Peer) and an auction owner (the auctioneer, e.g., EC-Operator). Since settlement is done based on physical measurements, sold and bought energy amounts may not be in balance and therefore the auction owner (EC-Operator) may generate some surplus.

This described marketplace approach has been fully implemented in Java as a stand-alone smart-contract and packaged as a Hyperledger Fabric chaincode. This is the central component of the FlexTrading DAPP.

7 Project Pilot

The goal of the pilot was to demonstrate real-time peer-to-peer (P2P) electricity trading between the members of energy community (EC), thus covering the two user stories US1 and US2 by combining the project's so-called Flexibility Management System (FMS, an aggregator platform), FlexCoins and FlexTrading. The FMS supported smart plugs and was extended with connectors and a trading agent for automatic bidding in the FlexTrading marketplace according to the real-time power measurements from these smart plugs.

7.1 Software Architecture

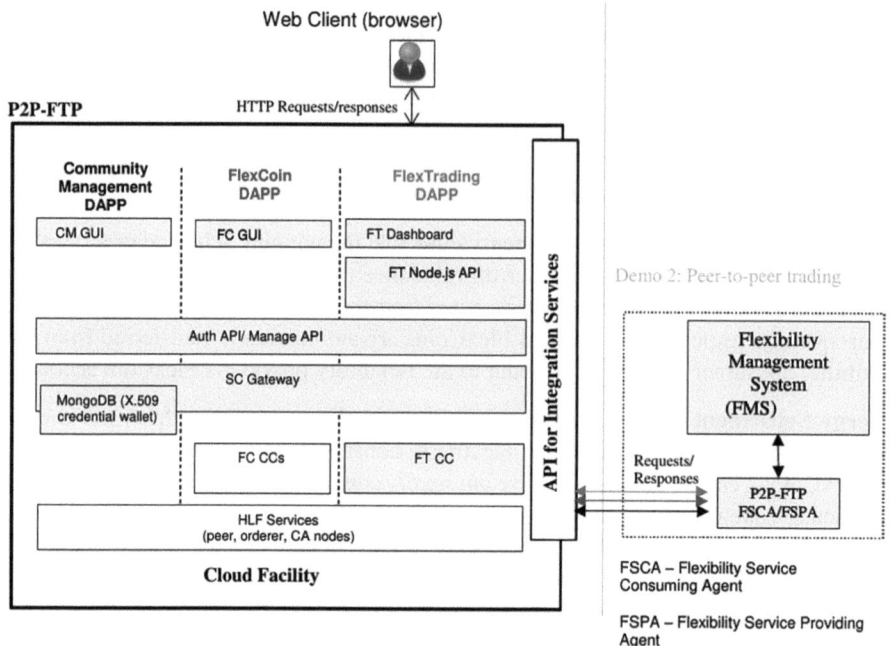

Fig. 10. Software architecture used for the P2P demonstrator

Figure 10 shows software architecture used for P2P demonstrator. The following main components are involved:

- **HLF services** are mandatory services, required to run a Hyperledger Fabric network and to deliver an immutable (blockchain) transaction ledger. HLF services include *peer, orderer, certificate authority* (CA) services.
- **FlexCoin Chaincode (FC CC) and FlexTrading Chaincode (FT CC)** encompass the business/governance logic of the FlexCoin and FlexTrading DAPPs, respectively. These chain codes run as stand-alone external services, connected to HLF peers. FT CC was integrated with FC CC so amounts of FlexCoins are automatically transferred between auction participants as a part of FlexTrading auction transaction processing.
- **Smart Contract Gateway (SC Gateway)** is a component, which allows applications (DAPPs) communicating with the underlying HLF services and smart contracts using standard HTT requests without the need to deal with sensitive user x.509 certificates. SC Gateway uses a MongoDB database to store P2P-FTP user x.509 certificates.
- **Auth/Manage APIs** provide application programming interfaces (APIs) to manage P2P-FTP user identities and to retrieve JSON Web Tokens (JWTs) required to access SC Gateway HTTP end-points.

- **FlexTrading Back-End API** (FT Node.js API) – which is a backend service developed using Node.js. It allows the FlexTrading DAPP dashboard (front-end) to communicate with the underlying FT CC and HLF network services, either directly or via SC Gateway.
- **Community Management (CM) GUI, FlexCoin (FC) GUI,** and **FlexTrading (FT) Admin Dashboard** are three independent front-ends / Graphical User Interfaces (GUIs) of Community Management, FlexCoin, and FlexTrading DAPPs, respectively. They allow various types of P2P-FTP end-users (e.g., EC Operators, EC Peers) interacting with P2P-FTP DAPPs using a standard Web client (e.g., Web-browser).

7.2 Demonstrator Deployment

P2P Demonstrator was deployed by 3 different organizations, namely ICOM, CERTH, and FLEXSHAPE:

- ICOM runs an instance of the Hyperledger Fabric (HLF) network by utilizing their Cloud Facility and CI/CD tools. For this, a production-ready instance of Kubernetes (K8S) is used, with a number of relevant HLF services (peers, orderer, certificate authority nodes, peer org admin cli) running in parallel. FlexTrading and FlexCoin (FC), FlexCoin Governance (GOV), and Hold Notary (NOT) smart-contracts are deployed as externalized chain codes. This means that actual smart contracts run externally to ICOM's HLF network by other organizations, namely FlexShape and CERTH.
- CERTH runs relevant FlexCoin (FC), FlexCoin Governance (GOV), and Hold Notary (NOT) chain code services, which together implement the complete FlexCoin application logic and communicates with ICOM's HLF peer for accessing and exchanging data. In addition, SC Gateway is deployed and made available to external applications. It translates incoming HTTP requests into HLF transactions and uses standard MongoDB for storing X.509 user certificates, required to grant users (user applications) access to the HLF network. In addition, user authorization and user management APIs as well as Community Management (CM) and FlexCoin (FC) Graphical User Interfaces (GUIs) are side deployed on the CERTH premises.
- FLEXSHAPE hosts components of P2P Demonstrator in their own Kubernetes (K8S) cluster. Here, the FlexTrading chaincode service implements the complete FlexTrading DAPP application logic and communicates with ICOM's HLF peer for accessing and exchanging relevant application data. Flexibility management system (FMS) is a sub-system (sub-set) of the FlexShape's Aggregator-as-a-Service platform [5]. It includes a number of relevant micro-services and applications. FlexTrading Market Service handles all communication with SC Gateway. In different deployment setups, it can as well communicate directly with the HLF network instance instead of via the SC Gateway, if this is required. The P2P Electricity Trading Admin APP is a GUI application, which allows (privileged) FMS users administering P2P auctions, approving/rejecting auction join and meter approval requests. Auction Runner component is responsible for clearing and settling P2P auctions periodically, serving as an external auction trigger for FlexTrading smart-contract. P2P Electricity Trading APP is an end-user application, which can be used by EC members to join auctions and approve meters. In addition, it allows creating automatic trading agents (robots), denoted as

P2P FSPA, which will trade electricity amounts, predicted according to real-time measurements from the TPLink Connector, which handles real-time communication with FMS-connected smart plugs.

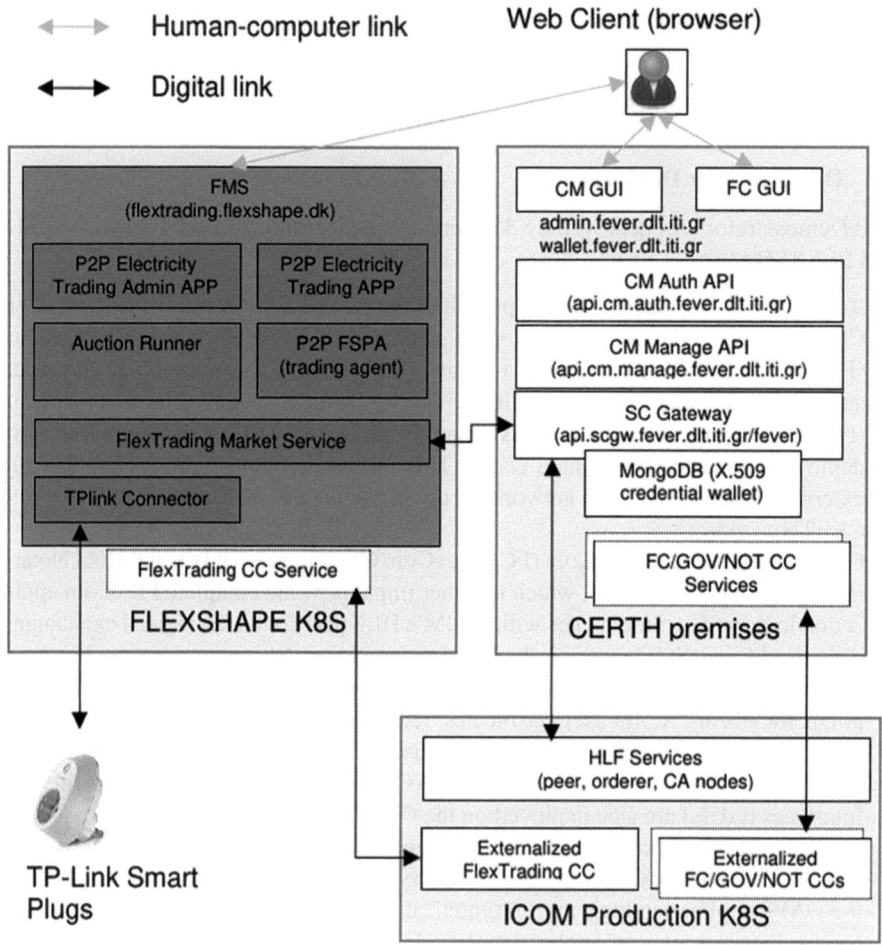

Fig. 11. P2P-FTP Demonstrator deployment

7.3 Pilot Phases

The pilot was realized through a number of semi-automatic actions performed by the admins/users, within the following three phases: preparation, demonstration, and post-demonstration.

Preparation Phase

1. Formed a group of physical users:
    a. 1 energy community operator (EC-Operator) – which will administer P2P electricity trading in an EC.

b. A number of electricity consumers (EC-Peers), with a (TPLink) smart plug connected to FMS Form a group of simulated users
   i. At least 1 electricity producer (EC-Peer), e.g., simulated PV panel, generator, battery, etc.
c. In Community Management (CM) GUI, create digital identities (accounts) for all physical and simulated users.
d. In FlexCoin GUI (Wallet), create FlexCoin accounts for all users from the user group; add initial deposit of FlexCoins for electricity buyers.
e. Configured FMS, using Prosumer Console, P2P Electricity Trading, and P2P Electricity Trading Admin, and other APPs:
   i. Permited access to FMS to all physical and simulated users; set relevant roles/permissions to these users.
   ii. Let EC-Operator to create a P2P auction, start the P2P auction runner.
   iii. Let all EC-Peers (physical and simulated users) to generate requests to join the P2P auction.
   iv. Let EC-Operator to approve the EC-Peer join requests.
   v. Let all EC-Peers to register their smart plugs, and request EC-Operator for their approval in FMS.
   vi. Let all EC-Peers create trading agents, linked with the same P2P auction and their individual smart plugs and trading strategies (incl., max el. Buying price, min el. Selling price);
   vii. Let all plug users connect some electrical device, which consumes some significant amount of electricity.
   viii. Activated simulated energy assets; made sure trading agents detects those simulated loads.

Demonstration Phase

1. Let the physical EC-Peers normally consume electricity using the device connected to the smart plugs.
2. When significant energy amounts are consumed, let the physical EC-Peers monitor automatic trading process in P2P Electricity Trading APP.
3. Let the physical EC-Operator monitor automatic trading process in P2P Electricity Trading Admin / System Modeller APPs.
4. Let EC-Peers monitor automatically generated transactions of FlexCoins in FlexCoin GUI.
5. Validated that the amounts of FlexCoins were transferred from electricity buyers (physical EC peers) to electricity sellers (simulated generators);

Post-demonstration Phase

1. Evaluated user experiences.
2. Extrapolated and concluded on the impact of the large-scale deployment of this solution.

In the preparation phase, the pilot required establishing a (virtual) EC community of electricity consumers and producers. Since FMS focuses on the integration of TPLink smart plugs only, no physical electricity producers are supported by FMS and thus included into this demo. As result, FMS-registered simulated electricity producers (generators) were used for this demo instead. Next, it was important to create digital user identities and FlexCoin accounts and add initial FlexCoin deposits into el. Consumer FlexCoin accounts to using the CM/FlexCoin DAPP GUIs. Then, it was important to configure FMS correctly: create a P2P auction, let EC-peers join that auction, add smart-plug devices into FMS and get approval from the EC-Operator on their inclusion into the P2P auction. Based on their individual preferences, each EC-Peer then had to configure a trading agent, which trades on the user behalf. Then, P2P electricity trading was able to start. EC-Peers simply used electricity with their smart plugs, the agent detected the consumption, made required predictions and automatically generated and submitted bids to the P2P auction. In addition, the agent periodically relayed measurements of consumed electricity from the internal meter of a smart plug via FMS to FlexTrading DAPP. During periodic settlement (triggered by the Auction Runner component), specific FlexCoin amounts were automatically transferred between accounts of auction participants. It was possible to monitor this in the FlexCoin DAPP GUI as well as P2P Electricity Trading (Admin) APP of FMS.

In the post-demonstration phase, it was important to evaluate (and learn from) user experiences, as well as extrapolate and conclude on the impact of the large-scale deployment of this solution.

8 Pilot Evaluation

8.1 Initial Experiment with the Complete Process

We first evaluated the Demonstrator in a small-scale experiment. The main purpose of this experiment was to validate that all relevant P2P Demonstrator components are operational, interoperable, and ready to be practically used in the German pilot described above. We started this experiment with a preparation please. For this experiment, a minimal test energy community (EC) was created. It consists of two users, an EC-Peer and an EC-Operator. The EC-Operator was created after the approval of another privileged user of CM DAPP. For the registration of the EC-Peer, the EC-Operator had to approve the generated EC-Peer registration request.

In the next step, FlexCoin accounts were created for both the EC-Peer and EC-Operator. These accounts initially have a balance of 0 FlexCoins. Before P2P trading can start, the electricity buyer (EC-Peer in this case) must deposit some FlexCoins into his/her FlexCoin account. This was accomplished using the FlexCoin GUI by generating a FlexCoin deposit request. Again, the EC-Operator had to approve (or reject) this request, e.g., after requesting some Euros or services in exchange to FlexCoins – see Fig. 12 below.

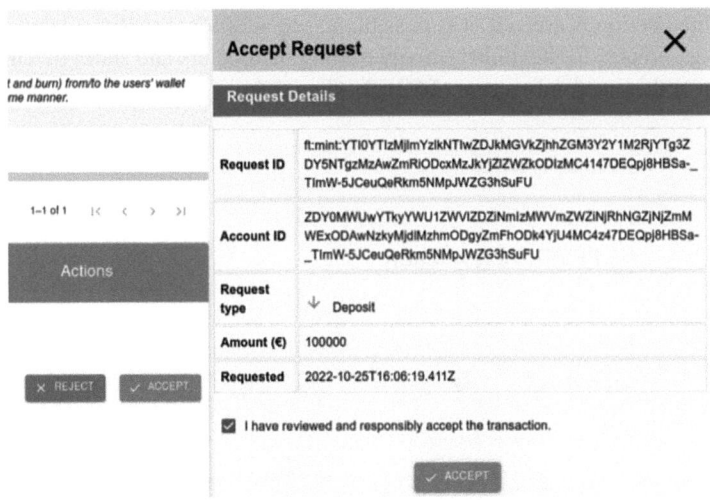

Fig. 12. FlexCoin GUI for FlexCoin deposit approval

After user registration was completed and initial FlexCoin deposits were made, the next step is the auction creation. A new P2P auction can be created using the P2P Trading Admin APP, only accessible by the EC-Operator. In the auction creation page, the EC-Operator has to specify several relevant auction parameters, including auction name, trading interval length, trading horizons, etc. – see Fig. 13. Key parameters include the auctioneer's (EC-Operator's) FlexCoin account and pricing mechanism, which determines the FlexCoin reserve account and pricing mechanism to use (e.g., price-as-bid, price-as-cleared).

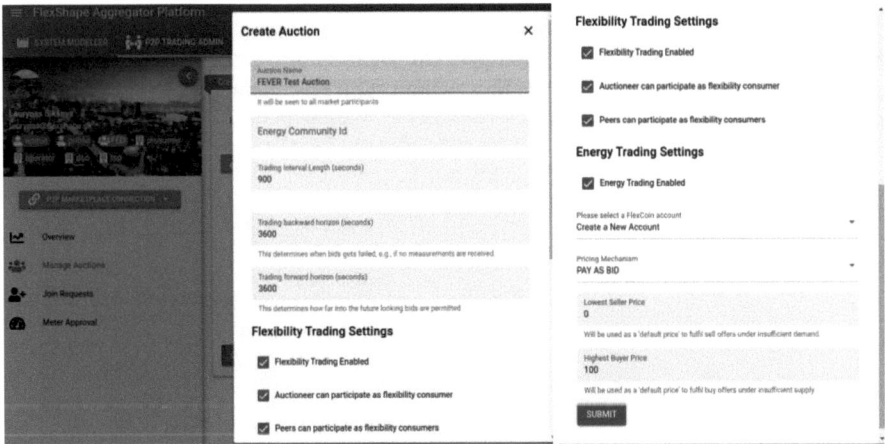

Fig. 13. P2P Trading Admin APP (FMS): New auction creation page

After the auction is created, it is possible to see additional runtime detail about the auction (e.g., creator Id, flexibility/energy trading parameters and state) by clicking the info button at the auction list page of P2P Trading Admin APP – see Fig. 14.

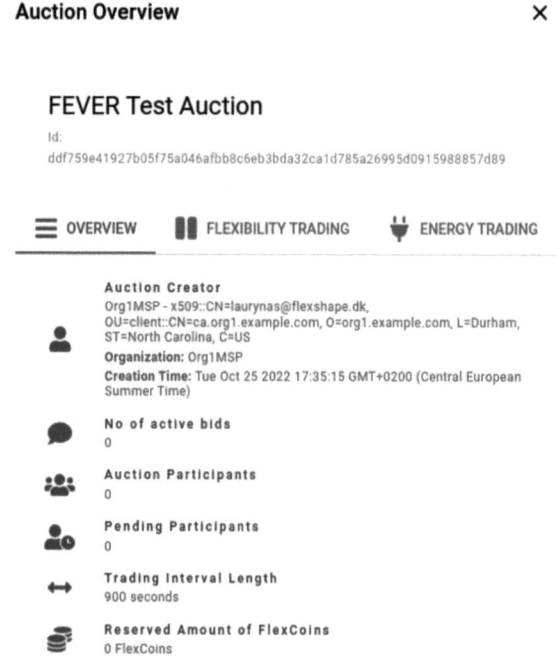

Fig. 14. P2P Trading Admin APP (FMS): Detail view of the auction

Finally, to make the newly created auction usable, a so-called auction runner has to be started to initiate an automatic and periodic clearing and settlement actions for the auction. This is done in the P2P Trading Admin APP by clicking the "play" button in the auction list page – see Fig. 15.

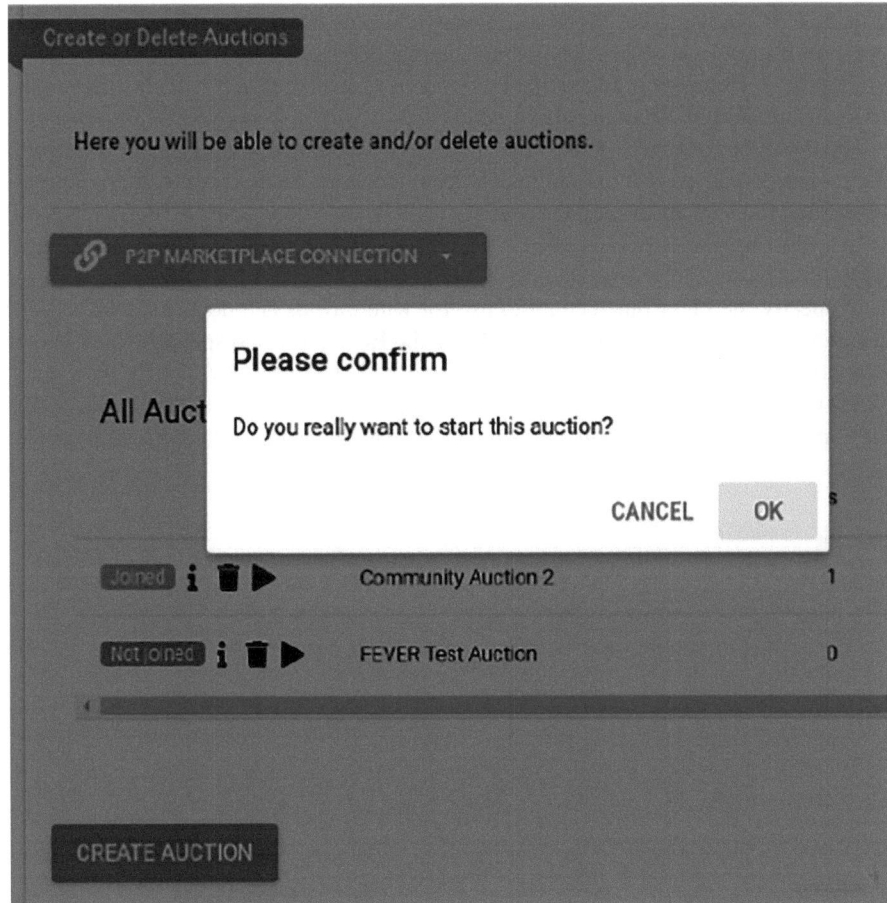

Fig. 15. P2P Trading Admin APP (FMS): Starting automatic clearing and settlement for the auction

Now, the auction is ready and can be used by (one of more) EC-Peer user(-s). The EC-Peer user can use the P2P auction using the P2P Trading APP of the FMS. It can be invoked directly from the list of APPs or through the Prosumer Console APP in FMS. On the first run, some configuration must be done by the EC-Peer user, as explained in the APP.

First, the EC-Peer joins the P2P auction by creating an auction join request specifying the auction and the relevant FlexCoin account. Next, the EC-Operator has to approve the auction join request of EC-Peer.

After the EC-Operator has approved the join request, the EC-Peer is able to see in the P2P Electricity Trading APP that he/she has successfully joined the auction.

The next step is to expose (add) relevant assets to the auction. This can be done by "adding" FMS- native energy assets, managed and/or pre-configured using other FMS APPs (e.g., Smart-Plug Loads). Thus, we added 1 of 2 configured smart-plug assets using the P2P Trading APP. These assets were added as smart meters, since smart plugs can report energy (and power) using their internal energy (and power) meters.

After the meter asset is added, the EC-Operator has to approve its inclusion into the auction using the P2P Trading Admin APP.

The final step for the EC-Peer is the creation of a so-called trading agent, which will collect energy (and power) measurements from the asset, make short term predictions, and trade in the P2P auction on the user's behalf. This is done using P2P Trading APP and by following a wizard. First, the EC-Peer has to give the trading agent a name. Next, the EC-Peer has to specify the auction in which the agent will trade. Then, the EC-Peer has to specify the FlexCoin account to use for withdrawals and deposits, which occur during the trading (settlement) process. Next, the EC-Peer has to specify which one of the approved assets the agent will monitor and will trade. Finally, the EC-Peer must specify the parameters of his/her preferred trading strategy, e.g., whether he/she wants to sell and/or buy electricity, and for what price. The final step of the wizard is seen in Fig. 16.

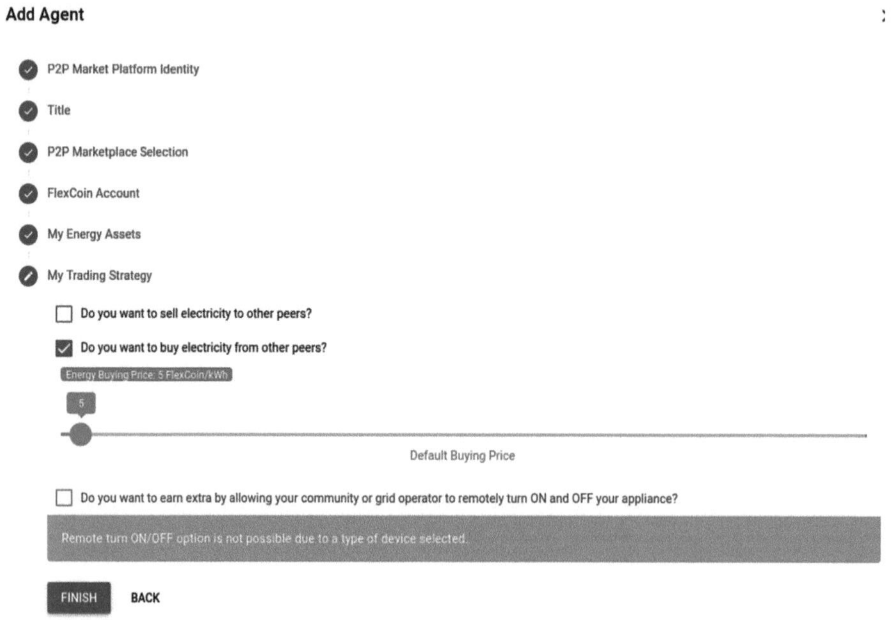

Fig. 16. P2P Trading APP (FMS): Specifying parameters of the trading strategy

After these steps, the agent is fully configured. Now, the trading agent is started. It continuously monitors the asset (smart plug) and the meter data uploaded onto the HLF network (using FlexTrading CC) and automatically generates P2P auction bids. When these bids get cleared or settled in the energy dispatch phase of the auction, trading results are retrieved from the P2P auction (using FlexTrading CC) and displayed to the user.

Figure 17 depicts the main dashboard of the P2P Trading APP, which shows in (near) real-time the metered, traded, and settled energy amounts in subsequent trading intervals, along with other relevant parameters (e.g., FlexCoin balance, energy bought today, etc.). In Fig. 17, we can see that in the time interval 6:15 to 6:30, 0.01874 kWh of electricity was bought at 100 FlexCoins/kWh – i.e., the pre-configured default buying price, used when no demand is satisfied (as in this case).

Fig. 17. P2P Trading APP (FMS): Trading status dashboard view

After keeping the system running for some time, we can witness the results of energy settlement, which takes place periodically but with a delay after each trading interval - after relevant (smart plug) meter data are uploaded onto the HLF network – see the details of the green banner in Fig. 18.

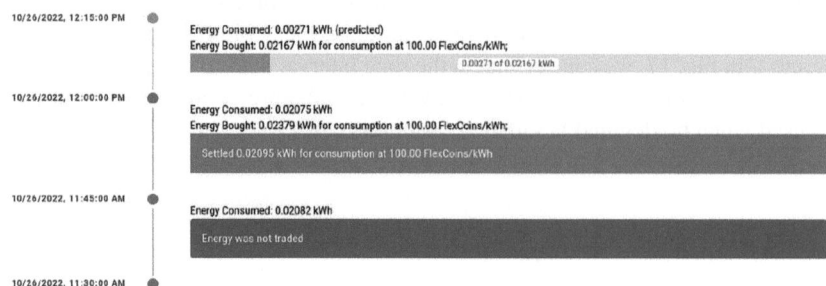

Fig. 18. Settlement Results

During the settlement, the system (FlexTrading CC) automatically transfers FlexCoin funds from the energy buyer's to the energy seller's (or the auctioneer's) FlexCoin account. We can validate that using the FlexCoin DAPP GUI – see Fig. 19.

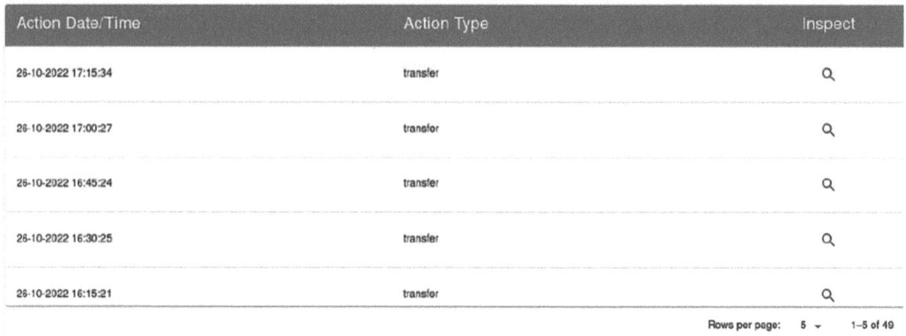

Action Date/Time	Action Type	Inspect
26-10-2022 17:15:34	transfer	🔍
26-10-2022 17:00:27	transfer	🔍
26-10-2022 16:45:24	transfer	🔍
26-10-2022 16:30:25	transfer	🔍
26-10-2022 16:15:21	transfer	🔍

Rows per page: 5 ▾   1–5 of 49

Fig. 19. FlexCoin DAPP GUI: A list of FlexCoin transactions, automatically generated during the auction settlement for the EC-Peer.

This is also reflected in the EC-Peer's FlexCoin account balance – see Fig. 20.

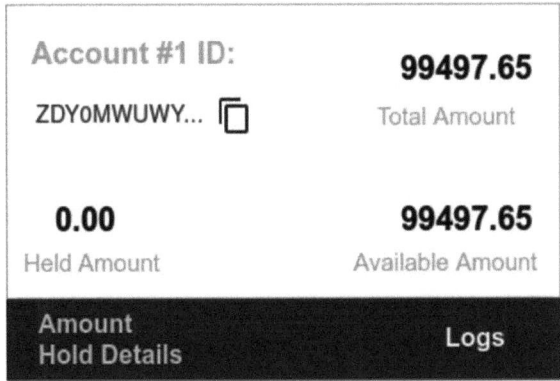

Fig. 20. FlexCoin DAPP GUI: The FlexCoin account balance after this experiment.

8.2 German Pilot Evaluation

Due to the lockdown, the German pilot ended up being smaller than planned. Ultimately, there was a community of 12 active smart-plug users (peers). Each peer bought electricity in FlexCoins within a single energy community based on actual power readings from 12 (of 21 available) smart plugs via FMS. A single energy community operator acted as the aggregator and sold electricity at a pre-defined fixed FlexCoin price to all these peers. Authorization, FlexCoin, and FlexTrading transactions were continuously generated and stored on the HLF.

We tested the following aspects of the P2P Demonstrator.

1. In the P2P Demonstrator, the interplay (/communication) between FMS (trading agent, market service, APPs) and P2P-FTP (CM, FlexCoin, FlexTrading DAPPs) works as expected.
2. Both the configuration and the demonstration phases can be realized using the P2P Demonstrator.
3. The internal P2P Demonstrator components work well in the end-to-end integrated setup:
   a. CM DAPP works well in the integration with FMS.
   b. The FMS trading agent can gather correct information from the smart plugs, generate and trade meaningful bids, and receive trading results.
   c. The FlexTrading smart contract can collect and process real-time measurements and bids from the trading agent, interact with FlexCoin eco-system.
   d. The P2P Electricity Trading APP and P2P Electricity Trading Admin APP of FMS can extract correct results from the FMS backend.

Our throughput measurements show that our current P2P-FTP deployment/setup used in FEVER offers transaction throughput rates, which are acceptable for real world applications. Here, user authorization transactions are fastest (7.38 transactions/sec). FlexCoin transactions are slightly slower (3.85 transactions/sec) due to involved HLF machinery (endorsing, etc.). FlexTrading transactions are slowest (1.08 transactions/sec) due to involved HLF machinery, and especially due to (internationally) distributed transaction processing using external chain code services (deployed in separate EU countries).

However, this performance is enough to support ECs with hundreds of members. Since the application requires 2 FlexTrading transactions (1 bid and 1 measurements) per 15 min per user to operate with an MTU of 15 min, our setup can scale to 15*60/2 = 450 users.

More details about the pilot evaluation can be found in FEVER Deliverable D7.3 [17].

9 Conclusion and Future Work

The rapidly increasing adoption of distributed electricity production, energy communities and flexible loads like heat pumps and EVs, creates a so far unmet need to trade energy and flexibility jointly instead of separately, which is what was done until now. This paper describes a solution for joint trading of electricity and flexibility within a local setting, e.g., energy communities, developed in the FEVER EU Horizon 2020 project. The solution is based on the FlexOffer concept and thus natively captures energy needs and their associated flexibilities in one joint data object. The paper explained the actors and procedures for trading and market clearing, the implementation based on the IBM HyperLedger Fabric blockchain platform and experimental results from a project pilot.

In future work, the marketplace will be matured and performance-tuned to support large-scale real-world deployment and ECs with many thousand members.

Acknowledgments. FEVER was funded by EC Horizon2020 under GA 864537.

Disclosure of Interests. The authors have no competing interests to declare that are relevant to the content of this article.

References

1. Pedersen, T.B., Siksnys, L., Neupane, B.: Modeling and Managing Energy Flexibility Using FlexOffers. IEEE SmartGridComm, ppp. 1–7 (2018)
2. FEVER Deliverable D5.1 - Interfaces and infrastructure specification of core blockchain services. https://fever-h2020.eu/data/deliverables/FEVER_D5.1_-_Interfaces_and_infrastructure_specification_of_core_blockchain_services.pdf
3. Siksnys, L., Pedersen, T.B.: Dependency-based FlexOffers: scalable management of flexible loads with dependencies. ACM e-Energy **11**(1–11), 13 (2016)
4. Copperplate model. https://www.researchgate.net/figure/Copper-Plate-Model-representation-of-the-system_fig2_241139992
5. Nord Pool Spot. https://www.nordpoolgroup.com/en/the-power-market/Day-ahead-market/
6. EPEX Spot https://www.epexspot.com/en
7. Radecke, J., Hefele, J., Hirth, L.: Markets for local flexibility in distribution networks. https://www.econstor.eu/handle/10419/204559
8. NODES Market. https://nodesmarket.com
9. Electron Market. https://electron.net

10. PLATONE Flexibility Market Platform. https://www.platone-h2020.eu/results/flexibility_market_platform
11. Powerledger market. https://powerledger.io
12. Grid Singularity. https://gridsingularity.com
13. Neupane, B., et al.: GOFLEX: extracting, aggregating and trading flexibility based on FlexOffers for 500+ prosumers in 3 European cities [operational systems paper]. e-Energy, 361–373 (2022)
14. The Nordic aFRR Capacity Markets. https://www.entsoe.eu/network_codes/eb/nordic-afrr-capacity-markets/
15. Frequency Containment Reserves. https://www.entsoe.eu/network_codes/eb/fcr/
16. FEVER Deliverable D7.3: Pilot validation report https://fever-h2020.eu/data/deliverables/FEVER_D7.3_Pilots-validation-report.pdf
17. Šikšnys, L., Khalefa, M.E., Pedersen, T.B.: Aggregating and Disaggregating Flexibility Objects. In: Ailamaki, A., Bowers, S. (eds.) SSDBM 2012. LNCS, vol. 7338, pp. 379–396. Springer, Heidelberg (2012). https://doi.org/10.1007/978-3-642-31235-9_25
18. Valsomatzis, E., Hose, K., Pedersen, T.B.: Balancing Energy Flexibilities Through Aggregation. DARE 2014, pp. 17–37 (2014)
19. Neupane, B., Pedersen, T.B., Thiesson, B.: Towards flexibility detection in device-level energy consumption. DARE 2014, pp. 1–16 (2014)
20. Neupane, B., Pedersen, T.B., Thiesson, B.: Evaluating the value of flexibility in energy regulation markets. e-Energy, 131–140 (2015)
21. Valsomatzis, E., Hose, K., Pedersen, T.B., Siksnys, L.: Measuring and Comparing Energy Flexibilities. In: EDBT/ICDT Workshops, pp. 78–85 (2015)
22. Siksnys, L., Valsomatzis, E., Hose, K., Pedersen, T.B.: Aggregating and disaggregating flexibility objects. IEEE Trans. Knowl. Data Eng. **27**(11), 2893–2906 (2015)
23. Valsomatzis, E., Pedersen, T.B., Abelló, A., Hose, K.: Aggregating energy flexibilities under constraints. SmartGridComm, pp. 484–490 (2016)
24. Valsomatzis, E., Pedersen, T.B., Abelló, A., Hose, K., Siksnys, L.: Towards constraint-based aggregation of energy flexibilities. e-Energy (Posters) 6:1–6:2 (2016)
25. Neupane, B., Siksnys, L., Pedersen, T.B.: Generation and evaluation of flex-offers from flexible electrical devices. e-Energy, 143–156 (2017). (Best Paper Award)
26. Valsomatzis, E., Pedersen, T.B., Abello, A.: Day-ahead trading of aggregated energy flexibility. e-Energy, 134–138 (2018)
27. Neupane, B., Pedersen, T.B., Thiesson, B.: Utilizing device-level demand forecasting for flexibility markets. e-Energy, pp. 108–118 (2018)
28. Frazzetto, D., Neupane, B., Pedersen, T.B., Nielsen, T.D.: Adaptive user-oriented direct load-control of residential flexible devices. e-Energy, 1–11 (2018)
29. Siksnys, L., Pedersen, T.B., Aftab, M., Neupane, B.: Flexibility modeling, management, and trading in bottom-up cellular energy systems. e-Energy, 170–180 (2019)
30. Lilliu, F., Pedersen, T.B., Siksnys, L.: Capturing battery flexibility in a general and scalable way using the FlexOffer model. SmartGridComm, 64–70 (2021)
31. Brusokas, J., Pedersen, T.B., Siksnys, L., Zhang, D., Chen, K.: HeatFlex: machine learning based data-driven flexibility prediction for individual heat pumps. e-Energy, 160–170 (2021)
32. Lilliu, F., Pedersen, T.B., Siksnys, L., Neupane, B.: Uncertain flexoffers, a scalable, uncertainty-aware model for energy flexibility. e-Energy, 448–449 (2022)
33. Lilliu, F., Pedersen, T.B., Siksnys, L.: Heat FlexOffers: a device-independent and scalable representation of electricity-heat flexibility. e-Energy, 374–385 (2023)
34. Khanal, S., Ho, N., Pedersen, T.B.: FDA-HeatFlex: scalable privacy-preserving temperature and flexibility prediction for heat pumps using federated domain adaptation. e-Energy, 172–183 (2023)

35. Lilliu, F., Pedersen, T.B., Siksnys, L., Neupane, B.: Uncertain FlexOffers: a scalable, uncertainty-aware model for energy flexibility. e-Energy, 30–41 (2023)
36. Lilliu, F., Brusokas, J., Pedersen, T.B.: Device-independent metrics for explicit, a priori measurement of energy flexibility. e-Energy (2024)
37. Lilliu, F., Laadhar, A., Thomsen, C., Recupero, D.R., Pedersen, T.B.: Extending the SAREF4ENER Ontology with Flexibility Based on FlexOffers. CoRR abs/2504.03595 (2025)
38. Siksnys, L., Pedersen, T.B.: Jointly trading energy and flexibility based on FlexOffers and blockchain. e-Energy, 996–997 (2025)
39. Brusokas, J., Lilliu, F., Tirupathi, S., Pedersen, T.B.: Selective-HeatFlex: selective forecasting of heat pump flexibility using model confidence. e-Energy, 191–207 (2025)
40. Liu, Z., Huang, B., Li, Y., Sun, Q., Pedersen, T.B., Wenzhong Gao, D.: Pricing game and blockchain for electricity data trading in low-carbon smart energy systems. IEEE Trans. Ind. Inform. **20**(4), 6446–6456 (2024)
41. Li, Y., et al.: Digital twin for secure peer-to-peer trading in cyber-physical energy systems. IEEE Trans. Netw. Sci. Eng. **12**(2), 669–683 (2025)

Author Index

A
Andersen, Jennie   93

B
Benna, Amel   45
Benslimane, Djamal   45
Burégio, Vanilson   45

C
Cazalens, Sylvie   93
Chikhaoui, Belkacem   45

E
Ezzeddine, Anna Bou   66

G
Gaignard, Alban   93
Gounaris, Anastasios   1

K
Kosmatopoulos, Andreas   1

L
Lamarre, Philippe   93

M
Maamar, Zakaria   45

P
Pavlík, Peter   66
Pedersen, Torben Bach   128

R
Rozinajová, Viera   66

S
Schleiss, Marc   66
Siksnys, Laurynas   128
Spitalas, Alexandros   1

T
Tsichlas, Kostas   1

MIX
Papier aus verantwortungsvollen Quellen
Paper from responsible sources
FSC® C105338

If you have any concerns about our products,
you can contact us on
ProductSafety@springernature.com

In case Publisher is established outside the EU,
the EU authorized representative is:
**Springer Nature Customer Service Center GmbH
Europaplatz 3, 69115 Heidelberg, Germany**

Printed by Libri Plureos GmbH
in Hamburg, Germany